Physical Metallurgy and Advanced Materials

Physical Metallurgy and Advanced Materials

Editor: Curtis Osborne

NY RESEARCH
P R E S S

New York

Published by NY Research Press
118-35 Queens Blvd., Suite 400,
Forest Hills, NY 11375, USA
www.nyresearchpress.com

Physical Metallurgy and Advanced Materials
Edited by Curtis Osborne

International Standard Book Number: 978-1-63238-601-4 (Hardback)

Cataloging-in-Publication Data

Physical metallurgy and advanced materials / edited by Curtis Osborne.
    p. cm.
Includes bibliographical references and index.
ISBN 978-1-63238-601-4
1. Physical metallurgy. 2. Materials--Technological innovations. 3. Metallurgy.
I. Osborne, Curtis.
TN690 .P49 2018
669.94--dc23

Contents

Preface

As physical metallurgy is one of the branches of metallurgy it deals with the thermal, electrical, magnetic and mechanical properties of metals and alloys. It includes applying the phase transformation elements and concepts to help understand the physical aspects of metals and alloys. The most common methodology used in this field is the CALPHAD. This book studies, analyses and upholds the pillars of physical metallurgy and its utmost significance in modern times. While understanding the long-term perspectives of the topics, it makes an effort in highlighting their impact as a modern tool for the growth of the discipline. The topics covered in the textbook offer the readers new insights in this field. It attempts to assist those with a goal of delving in this area.

A detailed account of the significant topics covered in this book is provided below:

Chapter 1- Metallurgy is a discipline of science that deals with the chemical and physical behaviour of metals, their compounds and alloys. Metals are ductile, lustrous and a few have magnetic qualities. Such features greatly boost its usage. This section is an overview of the subject matter incorporating all the major aspects of physical metallurgy.

Chapter 2- Thermal analysis studies changes in the properties of materials with a change in temperature. X-ray diffraction can reveal the finer details of a material as it has a smaller wavelength. Tools and techniques are an important component of any field of study. The following section elucidates the various tools and techniques that are related to physical metallurgy.

Chapter 3- The chapter deals with solidification of metals, the factors pertaining to its form, and its transformation and affecting factors, and also to deformation. It also explores cooling curve, a graph that represents shift in the state of matter. This chapter has been carefully written to provide an easy understanding of the varied facets of solidification and deformation.

Chapter 4- The detailed information of sequenced arrangement of atoms, ions or molecules in a crystalline material is called crystal structure. The crystal structure, however, feature many defects in them such as vacancy defects. When the atom is missing from a lattice side, it is called vacancy defect. The aspects elucidated in this section are of vital importance, and provide a better understanding of physical metallurgy.

I would like to make a special mention of my publisher who considered me worthy of this opportunity and also supported me throughout the process. I would also like to thank the editing team at the back-end who extended their help whenever required.

Editor

Understanding Physical Metallurgy and Nature of Metals

Metallurgy is a discipline of science that deals with the chemical and physical behaviour of metals, their compounds and alloys. Metals are ductile, lustrous and a few have magnetic qualities. Such features greatly boost its usage. This section is an overview of the subject matter incorporating all the major aspects of physical metallurgy.

Metallurgy

Metals and alloys have been in use since ages as jewelries, coins, tools, arms, and armors. Periods of human civilization have been named after metals & alloys viz. Copper Age / Bronze Age / Iron Age. However bulk use of metallic materials started with the commercial production of pig iron in blast furnace during the eighteenth century. Steel the most widely used metallic material became available in bulk only with the introduction of Bessemer Converters in mid nineteenth century (1856). Today it is one of the most extensively used engineering material. It is next to cement in terms of annual world production figures. Around 1.35 billion tons of crude steel is being produced in the world every year. India is the 5th largest producer of steel today. Our current production is around 75 Million ton. Selection of metals & alloys for a given application depends primarily on its properties, cost and availability. Materials are often not available in the form that could be directly put to use. It has to be extracted or synthesized from naturally occurring ingredients and subsequently processed to give it the desired shape and properties so that it is useful. Its cost has a direct correlation with the effort that is required to produce it. Energy spent in this process is a good indicator of such an effort. This is often referred to as embodied energy (stored energy). Table gives a list of common metals & materials along with their cost and energy required for their production from natural resources. It also includes the amount of CO_2 generated for every kg of the material being produced. Apart from energy this too gives a measure of the adverse impact a manmade material is likely to have on the environment. The amount of energy needed to produce aluminum alloys and the amount of CO_2 it produces are much higher than those for steel or concrete (a composite of cement sand and stone chips). The data in this table also suggests that there is a direct correlation between the cost of a material with either the magnitude of energy spent to produce it or the amount of CO_2 it emits during its production. No wonder that concrete and steel happen to be the cheapest materials of construction. Cu base alloys appear to an exception. High cost of raw materials could possibly a reason for this anomaly.

Table: - A comparison of energy needed to produce come of the common materials, their cost and their impact on environment measured in terms of CO_2 generated per unit of mass.

Material	Energy (MJ/kg)	Cost in $ /kg	CO_2 kg/kg
Al alloys	200-220	1.5-1.7	11.2-12.8
Polyethylene	75-78	1.3-1.5	2.0-2.2
Cu alloys	68-74	3.2-3.5	4.9-5.6
Steel	29-35	0.63 – 0.89	2.2-2.8
Glass	14-17	1.4-1.7	0.7-1.0
Concrete	1-1.3	0.041-0.062	0.13-0.15

Metallic materials are known for their ductility, electrical & thermal conductivity, strength, toughness and characteristic luster. A few of them have attractive magnetic properties as well. If you hit a piece of metal it gives a characteristic metallic sound. No wonder bells are made of metals (a special copper alloy). Apart from its ability to be cast in any desired shape and size; it is possible to form it by deformation and a variety of fabrication techniques.

Smelting gold in Nicaragua in the La Luz Gold Mine in Siuna and Bonanza about 1959.
Smelting is a basic step in obtaining usable quantities of most metals.

Pouring smelted gold into an ingot at the La Luz Gold Mine in Siuna, Nicaragua about 1959.

Metallurgy is a domain of materials science and engineering that studies the physical and chemical behavior of metallic elements, their inter-metallic compounds, and their mixtures, which are called alloys. Metallurgy is also the technology of metals: the way in which science is applied to the production of metals, and the engineering of metal components for usage in products for consumers and manufacturers. The production of metals involves the processing of ores to extract the metal they contain, and the mixture of metals, sometimes with other elements, to produce alloys. Metallurgy is distinguished from the craft of metalworking, although metalworking relies on metallurgy, as medicine relies on medical science, for technical advancement.

Metallurgy is subdivided into ferrous metallurgy (also known as *black metallurgy*) and non-ferrous metallurgy (also known as *colored metallurgy*).

Ferrous metallurgy involves processes and alloys based on iron while non-ferrous metallurgy involves processes and alloys based on other metals. The production of ferrous metals accounts for 95 percent of world metal production.

Etymology

The word was originally an alchemist's term for the extraction of metals from minerals, the ending *-urgy* signifying a process, especially manufacturing: it was discussed in this sense in the 1797 Encyclopædia Britannica. In the late 19th century it was extended to the more general scientific study of metals, alloys, and related processes.

In English, the pronunciation is the more common one in the UK and Commonwealth. The pronunciation is the more common one in the USA, and is the first-listed variant in various American dictionaries (e.g., *Merriam-Webster Collegiate*, *American Heritage*).

History

Gold headband from Thebes 750–700 BC

The earliest recorded metal employed by humans appears to be gold, which can be found free or "native". Small amounts of natural gold have been found in Spanish caves used during the late Paleolithic period, c. 40,000 BC. Silver, copper, tin and meteoric iron can also be found in native form, allowing a limited amount of metalworking in early cultures. Egyptian weapons made from meteoric iron in about 3000 BC were highly prized as "daggers from heaven".

Certain metals, notably tin, lead and (at a higher temperature) copper, can be recovered from their ores by simply heating the rocks in a fire or blast furnace, a process known as smelting. The first evidence of this extractive metallurgy dates from the 5th and 6th millennia BC and was found in the archaeological sites of Majdanpek, Yarmovac and Plocnik, all three in Serbia. To date, the earliest evidence of copper smelting is found at the Belovode site, including a copper axe from 5500 BC belonging to the Vinča culture. Other signs of early metals are found from the third millennium BC in places like Palmela (Portugal), Los Millares (Spain), and Stonehenge (United Kingdom). However, the ultimate beginnings cannot be clearly ascertained and new discoveries are both continuous and ongoing.

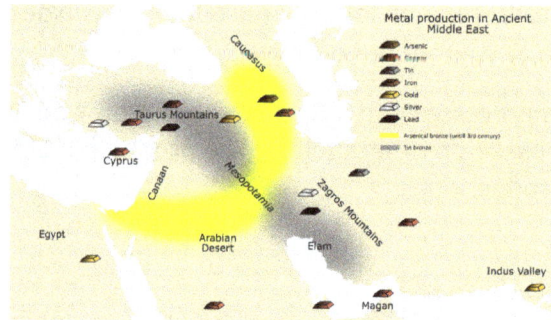

Mining areas of the ancient Middle East. Boxes colors: arsenic is in brown, copper in red, tin in grey, iron in reddish brown, gold in yellow, silver in white and lead in black. Yellow area stands for arsenic bronze, while grey area stands for tin bronze.

These first metals were single ones or as found. About 3500 BC, it was discovered that by combining copper and tin, a superior metal could be made, an alloy called bronze, representing a major technological shift known as the Bronze Age.

The extraction of iron from its ore into a workable metal is much more difficult than for copper or tin. The process appears to have been invented by the Hittites in about 1200 BC, beginning the Iron Age. The secret of extracting and working iron was a key factor in the success of the Philistines.

Historical developments in ferrous metallurgy can be found in a wide variety of past cultures and civilizations. This includes the ancient and medieval kingdoms and empires of the Middle East and Near East, ancient Iran, ancient Egypt, ancient Nubia, and Anatolia (Turkey), Ancient Nok, Carthage, the Greeks and Romans of ancient Europe, medieval Europe, ancient and medieval China, ancient and medieval India, ancient and medieval Japan, amongst others. Many applications, practices, and devices associated or involved in metallurgy were established in ancient China, such as the innovation of the blast furnace, cast iron, hydraulic-powered trip hammers, and double acting piston bellows.

A 16th century book by Georg Agricola called *De re metallica* describes the highly developed and complex processes of mining metal ores, metal extraction and metallurgy of the time. Agricola has been described as the "father of metallurgy".

Extraction

Furnace bellows operated by waterwheels, Yuan Dynasty, China.

Aluminium plant in Žiar nad Hronom (Central Slovakia)

Extractive metallurgy is the practice of removing valuable metals from an ore and refining the extracted raw metals into a purer form. In order to convert a metal oxide or sulphide to a purer metal, the ore must be reduced physically, chemically, or electrolytically.

Extractive metallurgists are interested in three primary streams: feed, concentrate (valuable metal oxide/sulphide), and tailings (waste). After mining, large pieces of the ore feed are broken through crushing and/or grinding in order to obtain particles small enough where each particle is either mostly valuable or mostly waste. Concentrating the particles of value in a form supporting separation enables the desired metal to be removed from waste products.

Mining may not be necessary if the ore body and physical environment are conducive to leaching. Leaching dissolves minerals in an ore body and results in an enriched solution. The solution is collected and processed to extract valuable metals.

Ore bodies often contain more than one valuable metal. Tailings of a previous process may be used as a feed in another process to extract a secondary product from the original ore. Additionally, a concentrate may contain more than one valuable metal. That concentrate would then be processed to separate the valuable metals into individual constituents.

Alloys

Casting bronze

Common engineering metals include aluminium, chromium, copper, iron, magnesium, nickel, titanium and zinc. These are most often used as alloys. Much effort has been placed on understanding the iron-carbon alloy system, which includes steels and cast irons. Plain carbon steels (those that contain essentially only carbon as an alloying element) are used in low-cost, high-strength applications where weight and corrosion are not a problem. Cast irons, including ductile iron, are also part of the iron-carbon system.

Stainless steel or galvanized steel are used where resistance to corrosion is important. Aluminium alloys and magnesium alloys are used for applications where strength and lightness are required.

Copper-nickel alloys (such as Monel) are used in highly corrosive environments and for non-magnetic applications. Nickel-based superalloys like Inconel are used in high-temperature applications such as gas turbines, turbochargers, pressure vessels, and heat exchangers. For extremely high temperatures, single crystal alloys are used to minimize creep.

Production

In production engineering, metallurgy is concerned with the production of metallic components for use in consumer or engineering products. This involves the production of alloys, the shaping, the heat treatment and the surface treatment of the product. The task of the metallurgist is to achieve balance between material properties such as cost, weight, strength, toughness, hardness, corrosion, fatigue resistance, and performance in temperature extremes. To achieve this goal, the operating environment must be carefully considered. In a saltwater environment, ferrous metals and some aluminium alloys corrode quickly. Metals exposed to cold or cryogenic conditions may endure a ductile to brittle transition and lose their toughness, becoming more brittle and prone to cracking. Metals under continual cyclic loading can suffer from metal fatigue. Metals under constant stress at elevated temperatures can creep.

Metalworking Processes

Metals are shaped by processes such as:

- casting – molten metal is poured into a shaped mold.
- forging – a red-hot billet is hammered into shape.
- rolling – a billet is passed through successively narrower rollers to create a sheet.
- laser cladding – metallic powder is blown through a movable laser beam (e.g. mounted on a NC 5-axis machine). The resulting melted metal reaches a substrate to form a melt pool. By moving the laser head, it is possible to stack the tracks and build up a three-dimensional piece.
- extrusion – a hot and malleable metal is forced under pressure through a die, which shapes it before it cools.
- sintering – a powdered metal is heated in a non-oxidizing environment after being compressed into a die.
- machining – lathes, milling machines, and drills cut the cold metal to shape.
- fabrication – sheets of metal are cut with guillotines or gas cutters and bent and welded into structural shape.

- 3D printing – Sintering or melting powder metal in a very small point on a moving 'print head' moving in 3D space to make any object to shape.

Cold-working processes, in which the product's shape is altered by rolling, fabrication or other processes while the product is cold, can increase the strength of the product by a process called work hardening. Work hardening creates microscopic defects in the metal, which resist further changes of shape.

Various forms of casting exist in industry and academia. These include sand casting, investment casting (also called thelost wax process), die casting, and continuous casting.

Heat Treatment

Metals can be heat-treated to alter the properties of strength, ductility, toughness, hardness and/or resistance to corrosion. Common heat treatment processes include annealing, precipitation strengthening, quenching, and tempering. The annealing process softens the metal by heating it and then allowing it to cool very slowly, which gets rid of stresses in the metal and makes the grain structure large and soft-edged so that when the metal is hit or stressed it dents or perhaps bends, rather than breaking; it is also easier to sand, grind, or cut annealed metal. Quenching is the process of cooling a high-carbon steel very quickly after heating, thus "freezing" the steel's molecules in the very hard martensite form, which makes the metal harder. There is a balance between hardness and toughness in any steel; the harder the steel, the less tough or impact-resistant it is, and the more impact-resistant it is, the less hard it is. Tempering relieves stresses in the metal that were caused by the hardening process; tempering makes the metal less hard while making it better able to sustain impacts without breaking.

Often, mechanical and thermal treatments are combined in what are known as thermo-mechanical treatments for better properties and more efficient processing of materials. These processes are common to high-alloy special steels, superalloys and titanium alloys.

Plating

Electroplating is a chemical surface-treatment technique. It involves bonding a thin layer of another metal such as gold, silver, chromium or zinc to the surface of the product. It is used to reduce corrosion as well as to improve the product's aesthetic appearance.

Thermal Spraying

Thermal spraying techniques are another popular finishing option, and often have better high temperature properties than electroplated coatings.

Microstructure

Metallurgists study the microscopic and macroscopic properties using metallography, a technique invented by Henry Clifton Sorby. In metallography, an alloy of interest is ground flat and polished to a mirror finish. The sample can then be etched to reveal the microstructure and macrostructure of the metal. The sample is then examined in an optical or electron microscope, and the image contrast provides details on the composition, mechanical properties, and processing history.

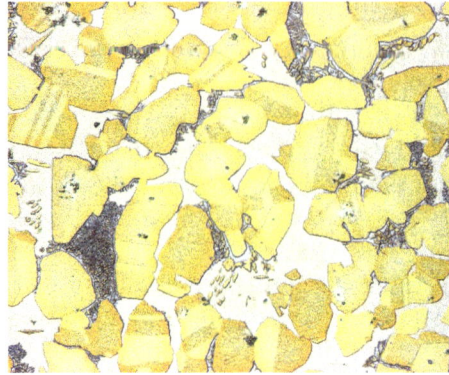

Metallography allows the metallurgist to study the microstructure of metals.

Crystallography, often using diffraction of x-rays or electrons, is another valuable tool available to the modern metallurgist. Crystallography allows identification of unknown materials and reveals the crystal structure of the sample. Quantitative crystallography can be used to calculate the amount of phases present as well as the degree of strain to which a sample has been subjected.

Conferences

EMC, the European Metallurgical Conference has developed to the most important networking business event dedicated to the non-ferrous metals industry in Europe. From the start of the conference sequence in 2001 at Friedrichshafen it was host of the most relevant metallurgists from all countries of the world. The European Metallurgical Conference is organized by GDMB Society of Metallurgists and Miners.

Physical Metallurgy

Physical metallurgy is one of the two main branches of the scientific approach to metallurgy, which considers in a systematic way the physical properties of metals and alloys. It is basically the fundamentals and applications of the theory of phase transformations in metal and alloys, as the title of classic, challenging monograph on the subject with this title. So, while chemical metallurgy involves the domain of reduction/oxidation of metals, physical metallurgy deals mainly with mechanical and magnetic/electric/thermal properties of metals – treated by the discipline of solid state physics. Calphad methodology, able to produce Phase diagrams which is the basis for evaluating or estimating physical properties of metals, relies on Computational thermodynamics i.e. on Chemical thermodynamics and could be considered a common and useful field for both the two sub-disciplines.

Atomic Structure

All solids are made of atoms which consist of subatomic particles; positively charged proton, neutron (neutral) and negatively charged electron. The entire mass of the atom is concentrated within a small nucleus consisting of proton and neutron whereas electrons having very little mass keep rotating about the nucleus in specific orbits. The sum total of proton and neutron gives the atomic mass number of an element; whereas the number of proton which is same as that of electron is its atomic number. The atom therefore is neutral. Hydrogen the lightest atom is made of one proton in the nucleus and an electron moving in a circular orbit known as k shell. The capacity of this orbit is 2. Helium is the next element with atomic number 2 and atomic mass 4. It has 2 protons & 2 neutrons in the nucleus and two electrons moving in a circular k shell in opposite directions. Since the capacity of this shell is 2 the outer orbit is full. Therefore even if two He atoms are brought close together there will be little interaction between the two. This is why He is a mono-atomic inert gas having an extremely low boiling point. However if two H atoms are brought close together it would try to form a bond by sharing the outer most electrons. This is why it is a diatomic gas. Atoms having higher atomic numbers would need more numbers of shells (K, L, M, N ....) and sub-shells (s, p, d, f ...) to accommodate orbiting electrons. The energy levels of these electrons are determined by four quantum numbers, principal quantum number (n) denoting shells, azimuthal quantum number (l) denoting sub-shells, magnetic quantum number (m) denoting effect of magnetic field on orbiting electrons, spin quantum number (ms) representing direction of rotation about its axis. The electrons are so arranged in different orbits that no two of these have the same quantum numbers. Following this rule it is possible to find out the capacity of each shell and sub-shell.

Stability of Atomic Structure

The stability of atomic structure depends on its tendency to interact with its neighbor. This is essentially determined by the number of electrons in its outer orbit often called as valence band with respect to its capacity. The capacity of K shell is 2. He having atomic number 2 has completely filled outer shell. Therefore its atomic structure is very stable. It is one of the most stable inert gases which exist in nature in elemental form. It's freezing and boiling points are extremely low. Apart from this all elements having 8 electrons in their valence band (outer orbit) have maximum stability with little chance for formation of bond. Inert gases (apart from He) like Ne, Ar, Kr, Xe, Rd have 8 electrons in their outer orbit. I am sure all of you are familiar with these concept introduced in your earlier courses. The genesis of bond formation in elements as they solidify lies in their atomic configuration. As long as the atoms are far apart there is hardly any interaction between these. However when two atoms are brought very close to each other the outer orbits would overlap. Since no two electrons can have the same quantum numbers additional sub-shells would have to form to accommodate more electrons. Depending on the way the outer shell electrons in neighboring atoms interact with each other there can be four different types of bonds in solids; metallic bond, ionic bond, covalent bond, and van der Waal bond. The properties of materials are determined by the nature of their bond. It also decides the type of crystal structure you expect. It is also possible to alter crystal structure by alloy addition and introduce defect structures thereby change properties of the material. Therefore it is important to know about atomic bond and crystal structure.

Example: Here are the electronic configuration of inert gases He $1s^2$; Ne $1s^2,2s^2,2p^6$; Ar $1s^2$ $2s^2$ $2p^6,3s^2,3p^6$. These are stable & have low melting point, He: 1K, Ne: 24K, Ar: 84K,..... Rn: 211K.

As atoms become larger the melting point increases. This indicates that the tendency to form bonds increases. This is because most of the sub-shells are not circular. The centre of all negatively charged electrons does not coincide with its nucleus where all the positive charges are concentrated. Therefore these tend to act as dipoles. This promotes formation of weak van der Waal bonds.

Example: Why is C so stable that it is found in nature in its atomic form? Its atomic configuration is: C $1s^2$ $2s^2$ $2p^2$. Note that the number of electron in its outer orbit (L shell) is 4. This is half filled.

Electrons in the outer orbit try to arrange themselves in such a fashion that its energy is the lowest (minimum). This happens if all the electrons have parallel spin. In such an event sp shell behaves as a hybridized orbit. This is known as Hund's rule. $1s^2, 2s^1 2sp^3$ (-px,-py,-pz) electronic configuration of Diamond. This is schematically shown as follows:

1s 2s 2sp 2sp 2sp

Metallic Bonding

Metallic bonding is a type of chemical bonding that arises from the electrostatic attractive force between conduction electrons (in the form of an electron cloud of delocalized electrons) and positively charged metal ions. It may be described as the sharing of *free* electrons among a lattice of positively charged ions (cations). Metallic bonding accounts for many physical properties of metals, such as strength, ductility, thermal and electrical resistivity and conductivity, opacity, and luster.

Metallic bonding is not the only type of chemical bonding a metal can exhibit, even as a pure substance. For example, elemental gallium consists of covalently-bound pairs of atoms in both liquid and solid state—these pairs form a crystal lattice with metallic bonding between them. Another example of a metal–metal covalent bond is mercurous ion (Hg_2^{2+}).

History

As chemistry developed into a science it became clear that metals formed the large majority of the periodic table of the elements and great progress was made in the description of the salts that can be formed in reactions with acids. With the advent of electrochemistry it became clear that metals generally go into solution as positively charged ions and the oxidation reactions of the metals became well understood in the electrochemical series. A picture emerged of metals as positive ions held together by an ocean of negative electrons.

With the advent of quantum mechanics this picture was given more formal interpretation in the form of the free electron model and its further extension, the nearly free electron model. In both of these models the electrons are seen as a gas traveling through the lattice of the solid with an energy that is essentially isotropic in that it depends on the square of the magnitude, *not* the direction of the momentum vector k. In three-dimensional k-space, the set of points of the highest filled levels (the Fermi surface) should therefore be a sphere. In the nearly free correction of the model, box-like Brillouin zones are added to k-space by the periodic potential experienced from the (ionic) lattice, thus mildly breaking the isotropy.

The advent of X-ray diffraction and thermal analysis made it possible to study the structure of crystalline solids, including metals and their alloys, and the construction of phase diagrams became accessible. Despite all this progress the nature of intermetallic compounds and alloys largely remained a mystery and their study was often empirical. Chemists generally steered away from anything that did not seem to follow Dalton's laws of multiple proportions and the problem was considered the domain of a different science, metallurgy.

The almost-free electron model was eagerly taken up by some researchers in this field, notably Hume-Rothery, in an attempt to explain why certain intermetallic alloys with certain compositions would form and others would not. Initially his attempts were quite successful. His idea was to add electrons to inflate the spherical Fermi-balloon inside the series of Brillouin-boxes and determine when a certain box would be full. This indeed predicted a fairly large number of observed alloy compositions. Unfortunately, as soon as cyclotron resonance became available and the shape of the balloon could be determined, it was found that the assumption that the balloon was spherical did not hold at all, except perhaps in the case of caesium. This reduced many of the conclusions to examples of how a model can sometimes give a whole series of correct predictions, yet still be wrong.

The free-electron debacle showed researchers that the model assuming that the ions were in a sea of free electrons needed modification, and so a number of quantum mechanical models such as band structure calculations based on molecular orbitals or the density functional theory were developed. In these models, one either departs from the atomic orbitals of neutral atoms that share their electrons or (in the case of density functional theory) departs from the total electron density. The free-electron picture has, nevertheless, remained a dominant one in education.

The electronic band structure model became a major focus not only for the study of metals but even more so for the study of semiconductors. Together with the electronic states, the vibrational states were also shown to form bands. Rudolf Peierls showed that, in the case of a one-dimensional row of metallic atoms, say hydrogen, an instability had to arise that would lead to the breakup of such a chain into individual molecules. This sparked an interest in the general question: When is collective metallic bonding stable and when will a more localized form of bonding take its place? Much research went into the study of clustering of metal atoms.

As powerful as the concept of the band structure proved to be in the description of metallic bonding, it does have a drawback. It remains a one-electron approximation to a multitudinous many-body problem. In other words, the energy states of each electron are described as if all the other electrons simply form a homogeneous background. Researchers like Mott and Hubbard realized that this was perhaps appropriate for strongly delocalized s- and p-electrons but for d-electrons, and even more for f-electrons the interaction with electrons (and atomic displacements) in the local environment may become stronger than the delocalization that leads to broad bands. Thus, the transition from localized unpaired electrons to itinerant ones partaking in metallic bonding became more comprehensible.

The Nature of Metallic Bonding

The combination of two phenomena gives rise to metallic bonding: delocalization of electrons and the availability of a far larger number of delocalized energy states than of delocalized electrons. The latter could be called electron deficiency.

In 2D

Graphene is an example of two-dimensional metallic bonding. Its metallic bonds are similar to aromatic bonding in benzene, naphthalene, anthracene, ovalene, and so on.

In 3D

Metal aromaticity in metal clusters is another example of delocalization, this time often in three-dimensional entities. Metals take the delocalization principle to its extreme and one could say that a crystal of a metal represents a single molecule over which all conduction electrons are delocalized in all three dimensions. This means that inside the metal one can generally not distinguish molecules, so that the metallic bonding is neither intra- nor intermolecular. 'Nonmolecular' would perhaps be a better term. Metallic bonding is mostly non-polar, because even in alloys there is little difference among the electronegativities of the atoms participating in the bonding interaction (and, in pure elemental metals, none at all). Thus, metallic bonding is an extremely delocalized communal form of covalent bonding. In a sense, metallic bonding is not a 'new' type of bonding at all, therefore, and it describes the bonding only as present in a *chunk* of condensed matter, be it crystalline solid, liquid, or even glass. Metallic vapors by contrast are often atomic (Hg) or at times contain molecules like Na_2 held together by a more conventional covalent bond. This is why it is not correct to speak of a single 'metallic bond'.

The delocalization is most pronounced for s- and p-electrons. For caesium it is so strong that the electrons are virtually free from the caesium atoms to form a gas constrained only by the surface of the metal. For caesium, therefore, the picture of Cs^+ ions held together by a negatively charged electron gas is not too inaccurate. For other elements the electrons are less free, in that they still experience the potential of the metal atoms, sometimes quite strongly. They require a more intricate quantum mechanical treatment (e.g., tight binding) in which the atoms are viewed as neutral, much like the carbon atoms in benzene. For d- and especially f-electrons the delocalization is not strong at all and this explains why these electrons are able to continue behaving as unpaired electrons that retain their spin, adding interesting magnetic properties to these metals.

Electron Deficiency and Mobility

Metal atoms contain few electrons in their valence shells relative to their periods or energy levels. They are electron deficient elements and the communal sharing does not change that. There remain far more available energy states than there are shared electrons. Both requirements for conductivity are therefore fulfilled: strong delocalization and partly filled energy bands. Such electrons can therefore easily change from one energy state into a slightly different one. Thus, not only do they become delocalized, forming a sea of electrons permeating the lattice, but they are also able to migrate through the lattice when an external electrical field is imposed, leading to electrical conductivity. Without the field, there are electrons moving equally in all directions. Under the field, some will adjust their state slightly, adopting a different wave vector. As a consequence, there will be more moving one way than the other and a net current will result.

The freedom of conduction electrons to migrate also give metal atoms, or layers of them, the capacity to slide past each other. Locally, bonds can easily be broken and replaced by new ones after the deformation. This process does not affect the communal metallic bonding very much. This gives rise to metals' typical characteristic phenomena of malleability and ductility. This is particularly true for pure elements. In the presence of dissolved impurities, the defects in the lattice that function as cleavage points may get blocked and the material becomes harder. Gold, for example, is very soft in pure form (24-karat), which is why alloys of 18-karat or lower are preferred in jewelry.

Metals are typically also good conductors of heat, but the conduction electrons only contribute partly to this phenomenon. Collective (i.e., delocalized) vibrations of the atoms known as phonons that travel through the solid as a wave, contribute strongly.

However, the latter also holds for a substance like diamond. It conducts heat quite well but *not* electricity. The latter is *not* a consequence of the fact that delocalization is absent in diamond, but simply that carbon is not electron deficient. The electron deficiency is an important point in distinguishing metallic from more conventional covalent bonding. Thus, we should amend the expression given above into: *Metallic bonding is an extremely delocalized communal form of electron deficient covalent bonding.*

Metallic Radius

Metallic radius is defined as one-half of the distance between the two adjacent metal ions in the metallic lattice. This radius depends on the nature of the atom as well as its environment—specifically, on the coordination number (CN), which in turn depends on the temperature and applied pressure.

When comparing periodic trends in the size of atoms it is often desirable to apply so-called Goldschmidt correction, which converts the radii to the values the atoms would have if they were 12-coordinated. Since metallic radii are always biggest for the highest coordination number, correction for less dense coordinations involves multiplying by x, where $0 < x < 1$. Specifically, for CN = 4, x = 0.88; for CN = 6, x = 0.96, and for CN = 8, x = 0.97. The correction is named after Victor Goldschmidt who obtained the numerical values quoted above.

The radii follow general periodic trends: they decrease across the period due to increase in the effective nuclear charge, which is not offset by the increased number of valence electrons. The radii also increase down the group due to increase in principal quantum number. Between rows 3 and 4, the lanthanide contraction is observed – there is very little increase of the radius down the group due to the presence of poorly shielding f orbitals.

Strength of the Bond

The atoms in metals have a strong attractive force between them. Much energy is required to overcome it. Therefore, metals often have high boiling points, with tungsten (5828 K) being extremely high. A remarkable exception are the elements of the zinc group: Zn, Cd, and Hg. Their electron configuration ends in ...ns^2 and this comes to resemble a noble gas configuration like that of helium more and more when going down in the periodic table because the energy distance to the empty np orbitals becomes larger. These metals are therefore relatively volatile, and are avoided in ultra-high vacuum systems.

Otherwise, metallic bonding can be very strong, even in molten metals, such as Gallium. Even though gallium will melt from the heat of one's hand just above room temperature, its boiling point is not far from that of copper. Molten gallium is therefore a very nonvolatile liquid thanks to its strong metallic bonding.

The strong bonding of metals in the liquid form demonstrates that the energy of a metallic bond is not a strong function of the direction of the metallic bond; this lack of bond directionality is a direct consequence of electron delocalization, and is best understood in contrast to the directional bonding of covalent bonds. The energy of a metallic bond is thus mostly a function of the amount of electrons which surround the metallic atom, as exemplified by the Embedded atom model. This typically results in metals assuming relatively simple, close-packed crystal structures, such as FCC, BCC, and HCP.

Given high enough cooling rates and appropriate alloy composition, metallic bonding can occur even in glasses with an amorphous structure.

Much biochemistry is mediated by the weak interaction of metal ions and biomolecules. Such interactions and their associated conformational change has been measured using dual polarisation interferometry.

Solubility and Compound Formation

Metals are insoluble in water or organic solvents unless they undergo a reaction with them. Typically this is an oxidation reaction that robs the metal atoms of their itinerant electrons, destroying the metallic bonding. However metals are often readily soluble in each other while retaining the metallic character of their bonding. Gold, for example, dissolves easily in mercury, even at room temperature. Even in solid metals, the solubility can be extensive. If the structures of the two metals are the same, there can even be complete solid solubility, as in the case of electrum, the alloys of silver and gold. At times, however, two metals will form alloys with different structures than either of the two parents. One could call these materials metal compounds, but, because materials with metallic bonding are typically not molecular, Dalton's law of integral proportions is not valid and often a range of stoichiometric ratios can be achieved. It is better to abandon such concepts as 'pure substance' or 'solute' is such cases and speak of phases instead. The study of such phases has traditionally been more the domain of metallurgy than of chemistry, although the two fields overlap considerably.

Localization and Clustering: from Bonding to Bonds

The metallic bonding in complicated compounds does not necessarily involve all constituent elements equally. It is quite possible to have an element or more that do not partake at all. One could picture the conduction electrons flowing around them like a river around an island or a big rock. It is possible to observe which elements do partake, e.g., by looking at the core levels in an X-ray photoelectron spectroscopy (XPS) spectrum. If an element partakes, its peaks tend to be skewed.

Some intermetallic materials e.g. do exhibit metal clusters, reminiscent of molecules and these compounds are more a topic of chemistry than of metallurgy. The formation of the clusters could be seen as a way to 'condense out' (localize) the electron deficient bonding into bonds of a more localized nature. Hydrogen is an extreme example of this form of condensation. At high pressures it is a metal. The core of the planet Jupiter could be said to be held together by a combination of metallic bonding and high pressure induced by gravity. At lower pressures however the bonding becomes entirely localized into a regular covalent bond. The localization is so complete that the (more familiar) H_2 gas results. A similar argument holds for an element like boron. Though it is electron deficient compared to carbon, it does not form a metal. Instead it has a number of complicated structures in which icosahedral B_{12} clusters dominate. Charge density waves are a related phenomenon.

As these phenomena involve the movement of the atoms towards or away from each other, they can be interpreted as the coupling between the electronic and the vibrational states (i.e. the phonons) of the material. A different such electron-phonon interaction is thought to cause a very different result at low temperatures, that of superconductivity. Rather than blocking the mobility of the charge carriers by forming electron pairs in localized bonds, Cooper-pairs are formed that no longer experience any resistance to their mobility.

Optical Properties

The presence of an ocean of mobile charge carriers has profound effects on the optical properties of metals. They can only be understood by considering the electrons as a *collective* rather than considering the states of individual electrons involved in more conventional covalent bonds.

Light consists of a combination of an electrical and a magnetic field. The electrical field is usually able to excite an elastic response from the electrons involved in the metallic bonding. The result is that photons are not able to penetrate very far into the metal and are typically reflected. They bounce off, although some may also be absorbed. This holds equally for all photons of the visible spectrum, which is why metals are often silvery white or grayish with the characteristic specular reflection of metallic luster. The balance between reflection and absorption determines how white or how gray they are, although surface tarnish can obscure such observations. Silver, a very good metal with high conductivity is one of the whitest.

Notable exceptions are reddish copper and yellowish gold. The reason for their color is that there is an upper limit to the frequency of the light that metallic electrons can readily respond to, the plasmon frequency. At the plasmon frequency, the frequency-dependent dielectric function of the free electron gas goes from negative (reflecting) to positive (transmitting); higher frequency photons are not reflected at the surface, and do not contribute to the color of the metal. There are some materials like indium tin oxide (ITO) that are metallic conductors (actually degenerate semiconductors) for which this threshold is in the infrared, which is why they are transparent in the visible, but good mirrors in the IR.

For silver the limiting frequency is in the far UV, but for copper and gold it is closer to the visible. This explains the colors of these two metals. At the surface of a metal resonance effects known as surface plasmons can result. They are collective oscillations of the conduction electrons like a ripple in the electronic ocean. However, even if photons have enough energy they usually do not have enough momentum to set the ripple in motion. Therefore, plasmons are hard to excite on a bulk metal. This is why gold and copper still look like lustrous metals albeit with a dash of color. However, in colloidal gold the metallic bonding is confined to a tiny metallic particle, preventing the oscillation wave of the plasmon from 'running away'. The momentum selection rule is therefore broken, and the plasmon resonance causes an extremely intense absorption in the green with a resulting beautiful purple-red color. Such colors are orders of magnitude more intense than ordinary absorptions seen in dyes and the like that involve individual electrons and their energy states.

When electrons in valence band are equally shared by all atoms it results in metallic bond. The range over which atomic forces are significant is very small. When two atoms are brought close enough the outer orbit or the valence band start overlapping, the force of attraction becomes significant. Since no two electrons can have identical set of quantum numbers the valence band must split to accommodate more electrons.

This type of distribution makes metallic bond non-directional. The strength of the bond depends on the number of valence electrons per atom, as if the electrons play the role of a glue to join layers of atoms. Look at the atomic configuration of elements in the fourth row of the periodic table. There are several transition elements (where the number of electrons in the outer most shell remains unchanged till the lower orbits are filled up) beyond Ca in this row. The inner orbits of Ca are similar to that of Ar. In addition it has 2 electrons in 4s. Its configuration is represented as Ca [Ar] $4s^2$. Electrons in atoms of subsequent elements tend to occupy 3d shell. For example the configuration of Ti is [Ar] $3d^2 4s^2$. This trend continues till Zn: [Ar] $^3d^{10} 4s^2$ when 3d shell gets completely filled up. The valence band in these elements is made up of the electrons in overlapping 3d & 4s shells. Electrons having unpaired spin belong to the valence band. The number of such electrons per atom increases along the period (row). It reaches its peak at Cr and there after it decreases. This is the reason why the melting point (which is an indicator of the strength of the bond that develops between atoms) of the elements along this row keeps increasing reaches its peak at Cr and then decreases.

Since in metals electrons are shared by all atoms and the electrons occupying the top most energy level in valence band (known as Fermi level) could easily move to the vacant higher energy levels; they have high thermal & electrical conductivity, moderately high elastic modulus and are amenable to plastic deformation (ductile).

Ionic Bonding

Sodium and fluorine undergoing a redox reaction to form sodium fluoride. Sodium loses its outer electron to give it a stable electron configuration, and this electron enters the fluorine atom exothermically. The oppositely charged ions – typically a great many of them – are then attracted to each other to form a solid.

Ionic bonding is a type of chemical bond that involves the electrostatic attraction between oppositely charged ions, and is the primary interaction occurring in ionic compounds. The ions are atoms that have gained one or more electrons (known as anions, which are negatively charged) and atoms that have lost one or more electrons (known as cations, which are positively charged). This transfer of electrons is known as electrovalence in contrast to covalence. In the simplest case, the cation is a metal atom and the anion is a nonmetal atom, but these ions can be of a more complex nature, e.g. molecular ions like NH_4^+ or SO_4^{2-}. In simpler words, an ionic bond is the transfer of electrons from a metal to a non-metal in order to obtain a full valence shell for both atoms.

It is important to recognize that *clean* ionic bonding – in which one atom or molecule completely share an electron from another – cannot exist: all ionic compounds have some degree of covalent bonding, or electron sharing. Thus, the term "ionic bonding" is given when the ionic character is greater than the covalent character – that is, a bond in which a large electronegativity difference exists between the two atoms, causing the bonding to be more polar (ionic) than in covalent bonding where electrons are shared more equally. Bonds with partially ionic and partially covalent character are called polar covalent bonds.

Ionic compounds conduct electricity when molten or in solution, typically as a solid. Ionic compounds generally have a high melting point, depending on the charge of the ions they consist of. The higher the charges the stronger the cohesive forces and the higher the melting point. They also tend to be soluble in water. Here, the opposite trend roughly holds: the weaker the cohesive forces, the greater the solubility.

Overview

Atoms that have an almost full or almost empty valence shell tend to be very reactive. Atoms that are strongly electronegative (as is the case with halogens) often have only one or two empty orbitals in their valence shell, and frequently bond with other molecules or gain electrons to form anions. Atoms that are weakly electronegative (such as alkali metals) have relatively few valence electrons that can easily be shared with atoms that are strongly electronegative. As a result, weakly electronegative atoms tend to distort their electrons cloud and form cations.

Formation

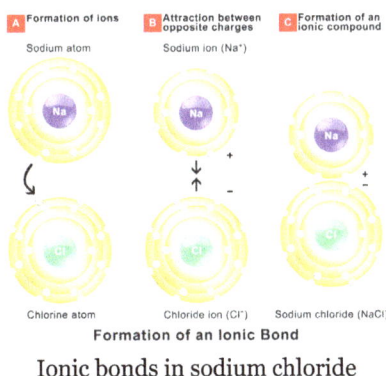

Formation of an Ionic Bond

Ionic bonds in sodium chloride

Ionic bonding can result from a redox reaction when atoms of an element (usually metal), whose ionization energy is low, give some of their electrons to achieve a stable electron configuration. In doing so, cations are formed. The atom of another element (usually nonmetal), whose electron affinity is positive, then accepts the electron(s), again to attain a stable electron configuration, and after accepting electron(s) the atom becomes an anion. Typically, the stable electron configuration is one of the noble gases for elements in the s-block and the p-block, and particular stable electron configurations for d-block and f-block elements. The electrostatic attraction between the anions and cations leads to the formation of a solid with a crystallographic lattice in which the ions are stacked in an alternating fashion. In such a lattice, it is usually not possible to distinguish discrete molecular units, so that the compounds formed are not molecular in nature. However, the ions themselves can be complex and form molecular ions like the acetate anion or the ammonium cation.

For example, common table salt is sodium chloride. When sodium (Na) and chlorine (Cl) are combined, the sodium atoms each lose an electron, forming cations (Na^+), and the chlorine atoms each gain an electron to form anions (Cl^-). These ions are then attracted to each other in a 1:1 ratio to form sodium chloride (NaCl).

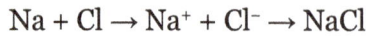

$$Na + Cl \rightarrow Na^+ + Cl^- \rightarrow NaCl$$

However, to maintain charge neutrality, strict ratios between anions and cations are observed so that ionic compounds, in general, obey the rules of stoichiometry despite not being molecular compounds. For compounds that are transitional to the alloys and possess mixed ionic and metallic bonding, this may not be the case anymore. Many sulfides, e.g., do form non-stoichiometric compounds.

Many ionic compounds are referred to as salts as they can also be formed by the neutralization reaction of an Arrhenius base like NaOH with an Arrhenius acid like HCl

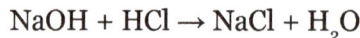

$$NaOH + HCl \rightarrow NaCl + H_2O$$

The salt NaCl is then said to consist of the acid rest Cl^- and the base rest Na^+.

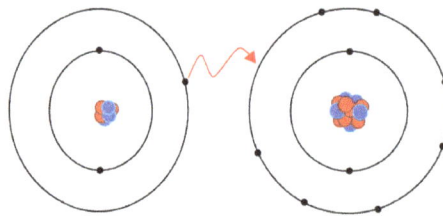

Representation of ionic bonding between lithium and fluorine to form lithium fluoride. Lithium has a low ionization energy and readily gives up its lone valence electron to a fluorine atom, which has a positive electron affinity and accepts the electron that was donated by the lithium atom. The end-result is that lithium is isoelectronic with helium and fluorine is isoelectronic with neon. Electrostatic interaction occurs between the two resulting ions, but typically aggregation is not limited to two of them. Instead, aggregation into a whole lattice held together by ionic bonding is the result.

The removal of electrons from the cation is endothermic, raising the system's overall energy. There may also be energy changes associated with breaking of existing bonds or the addition of more than one electron to form anions. However, the action of the anion's accepting the cation's valence electrons and the subsequent attraction of the ions to each other releases (lattice) energy and, thus, lowers the overall energy of the system.

Ionic bonding will occur only if the overall energy change for the reaction is favorable. In general, the reaction is exothermic, but, e.g., the formation of mercuric oxide (HgO) is endothermic. The charge of the resulting ions is a major factor in the strength of ionic bonding, e.g. a salt C^+A^- is held together by electrostatic forces roughly four times weaker than $C^{2+}A^{2-}$ according to Coulombs law, where C and A represent a generic cation and anion respectively. Of course the sizes of the ions and the particular packing of the lattice are ignored in this simple argument.

Structures

Ionic compounds in the solid state form lattice structures. The two principal factors in determining the form of the lattice are the relative charges of the ions and their relative sizes. Some structures are adopted by a number of compounds; for example, the structure of the rock salt sodium chloride is also adopted by many alkali halides, and binary oxides such as MgO. Pauling's rules provide guidelines for predicting and rationalizing the crystal structures of ionic crystals.

Bond Strength

For a solid crystalline ionic compound the enthalpy change in forming the solid from gaseous ions is termed the lattice energy. The experimental value for the lattice energy can be determined using the Born-Haber cycle. It can also be calculated (predicted) using the Born-Landé equation as the sum of the electrostatic potential energy, calculated by summing interactions between cations and anions, and a short-range repulsive potential energy term. The electrostatic potential can be expressed in terms of the inter-ionic separation and a constant (Madelung constant) that takes account of the geometry of the crystal. The further away from the nucleus the weaker the shield. The Born-Landé equation gives a reasonable fit to the lattice energy of, e.g., sodium chloride, where the calculated (predicted) value is −756 kJ/mol, which compares to −787 kJ/mol using the Born-Haber cycle.

Polarization Effects

Ions in crystal lattices of purely ionic compounds are spherical; however, if the positive ion is small and/or highly charged, it will distort the electron cloud of the negative ion, an effect summarised in Fajans' rules. This polarization of the negative ion leads to a build-up of extra charge density between the two nuclei, i.e., to partial covalency. Larger negative ions are more easily polarized, but the effect is usually important only when positive ions with charges of 3+ (e.g., Al^{3+}) are involved. However, 2+ ions (Be^{2+}) or even 1+ (Li^+) show some polarizing power because their sizes are so small (e.g., LiI is ionic but has some covalent bonding present). Note that this is not the ionic polarization effect that refers to displacement of ions in the lattice due to the application of an electric field.

Comparison with Covalent Bonding

In ionic bonding, the atoms are bound by attraction of opposite ions, whereas, in covalent bonding, atoms are bound by sharing electrons to attain stable electron configurations. In covalent bonding, the molecular geometry around each atom is determined by valence shell electron pair repulsion VSEPR rules, whereas, in ionic materials, the geometry follows maximum packing rules. One could say that covalent bonding is more *directional* in the sense that the energy penalty for not adhering to the optimum bond angles is large, whereas ionic bonding has no such penalty. There are no shared electron pairs to repel each other, the ions should simply be packed as efficiently as possible. This often leads to much higher coordination numbers. In NaCl, each ion has 6 bonds and all bond angles are 90 degrees. In CsCl the coordination number is 8. By comparison carbon typically has a maximum of four bonds.

Purely ionic bonding cannot exist, as the proximity of the entities involved in the bonding allows some degree of sharing electron density between them. Therefore, all ionic bonding has some covalent character. Thus, bonding is considered ionic where the ionic character is greater than the covalent character. The larger the difference in electronegativity between the two types of atoms involved in the bonding, the more ionic (polar) it is. Bonds with partially ionic and partially covalent character are called polar covalent bonds. For example, Na–Cl and Mg–O interactions have a few percent covalency, while Si–O bonds are usually ~50% ionic and ~50% covalent. Pauling estimated that an electronegativity difference of 1.7 (on the Pauling scale) corresponds to 50% ionic character, so that a difference greater than 50% corresponds to a bond which is predominantly ionic. Ionic character in covalent bonds can be directly measured for atoms having quadrupolar nuclei (^{2}H, ^{14}N, ^{81,79}Br, ^{35,37}Cl or ^{127}I). These nuclei are generally objects of NQR nuclear quadrupole resonance and NMR nuclear magnetic resonance studies. Interactions between the nuclear quadrupole moments Q and the electric field gradients (EFG) are characterized via the nuclear quadrupole coupling constants QCC = $e^2q_{zz}Q/h$ where the eq_{zz} term corresponds to the principal component of the EFG tensor and e is the elementary charge. In turn, the electric field gradient opens the way to description of bonding modes in molecules when the QCC values are accurately determined by NMR or NQR methods.

In general, when ionic bonding occurs in the solid (or liquid) state, it is not possible to talk about a single "ionic bond" between two individual atoms, because the cohesive forces that keep the lattice together are of a more collective nature. This is quite different in the case of covalent bonding, where we can often speak of a distinct bond localized between two particular atoms. However, even if ionic bonding is combined with some covalency, the result is *not* necessarily discrete bonds of a localized character. In such cases, the resulting bonding often requires description in terms of a band structure consisting of gigantic molecular orbitals spanning the entire crystal. Thus, the bonding in the solid often retains its collective rather than localized nature. When the difference in electronegativity is decreased, the bonding may then lead to a semiconductor, a semimetal or eventually a metallic conductor with metallic bonding.

Ionic bond develops between two atoms having widely different electro-negativity. For example such a bond can form between Na and Cl; two elements: one in group I and the other in group VII. Na has one loosely bound electron in its outer shell whereas Cl is short of one electron in its outer orbit to reach the stable octet configuration. Therefore if these two atoms come closer Na would lose one outer electron to form a stable octet and Cl would pick it up to attain a stable structure. Consequently Na & Cl atoms become charged particles (ions). The bond that forms between the two is due to the force of attraction between the two oppositely charged ions. Since the electric field is uniformly distributed in all direction such bonds do not exhibit any directionality. Like metallic bond ionic bond is also strong. Materials having such a bond have high elastic modulus. The electrons are tightly bound and are not free to move. Therefore its thermal & electrical conductivity are poor. However such materials exhibit ionic conductivity if molten or if present in aqueous solution. Ionic materials are brittle with little ductility.

Covalent Bond

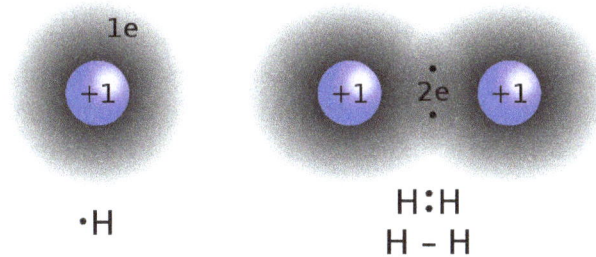

A covalent bond forming H_2 (right) where two hydrogen atoms share the two electrons

A covalent bond, also called a molecular bond, is a chemical bond that involves the sharing of electron pairs between atoms. These electron pairs are known as shared pairs or bonding pairs, and the stable balance of attractive and repulsive forces between atoms, when they share electrons, is known as covalent bonding. For many molecules, the sharing of electrons allows each atom to attain the equivalent of a full outer shell, corresponding to a stable electronic configuration.

Covalent bonding includes many kinds of interactions, including σ-bonding, π-bonding, metal-to-metal bonding, agostic interactions, bent bonds, and three-center two-electron bonds. The term *covalent bond* dates from 1939. The prefix *co-* means *jointly, associated in action, partnered to a lesser degree,* etc.; thus a "co-valent bond", in essence, means that the atoms share "valence", such as is discussed in valence bond theory.

In the molecule H_2, the hydrogen atoms share the two electrons via covalent bonding. Covalency is greatest between atoms of similar electronegativities. Thus, covalent bonding does not necessarily require that the two atoms be of the same elements, only that they be of comparable electronegativity. Covalent bonding that entails sharing of electrons over more than two atoms is said to be delocalized.

History

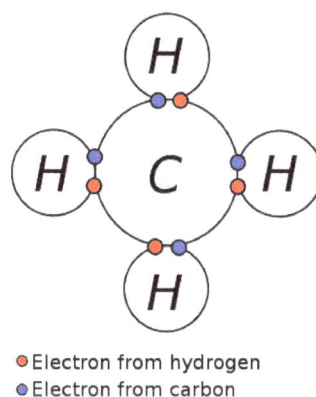

● Electron from hydrogen
● Electron from carbon

Early concepts in covalent bonding arose from this kind of image of the molecule of methane. Covalent bonding is implied in the Lewis structure by indicating electrons shared between atoms.

The term *covalence* in regard to bonding was first used in 1919 by Irving Langmuir in a *Journal of the American Chemical Society* article entitled "The Arrangement of Electrons in Atoms and Molecules". Langmuir wrote that "we shall denote by the term *covalence* the number of pairs of electrons that a given atom shares with its neighbors."

The idea of covalent bonding can be traced several years before 1919 to Gilbert N. Lewis, who in 1916 described the sharing of electron pairs between atoms. He introduced the *Lewis notation* or *electron dot notation* or *Lewis dot structure*, in which valence electrons (those in the outer shell) are represented as dots around the atomic symbols. Pairs of electrons located between atoms represent covalent bonds. Multiple pairs represent multiple bonds, such as double bonds and triple bonds. An alternative form of representation, has bond-forming electron pairs represented as solid lines.

Lewis proposed that an atom forms enough covalent bonds to form a full (or closed) outer electron shell. The carbon atom has a valence of four and is, therefore, surrounded by eight electrons (the octet rule), four from the carbon itself and four from the hydrogens bonded to it. Each hydrogen has a valence of one and is surrounded by two electrons (a duet rule) – its own one electron plus one from the carbon. The numbers of electrons correspond to full shells in the quantum theory of the atom; the outer shell of a carbon atom is the $n = 2$ shell, which can hold eight electrons, whereas the outer (and only) shell of a hydrogen atom is the $n = 1$ shell, which can hold only two.

While the idea of shared electron pairs provides an effective qualitative picture of covalent bonding, quantum mechanics is needed to understand the nature of these bonds and predict the structures and properties of simple molecules. Walter Heitler and Fritz London are credited with the first successful quantum mechanical explanation of a chemical bond (molecular hydrogen) in 1927. Their work was based on the valence bond model, which assumes that a chemical bond is formed when there is good overlap between the atomic orbitals of participating atoms.

Types of Covalent Bonds

Atomic orbitals (except for s orbitals) have specific directional properties leading to different types of covalent bonds. Sigma (σ) bonds are the strongest covalent bonds and are due to head-on overlapping of orbitals on two different atoms. A single bond is usually a σ bond. Pi (π) bonds are weaker and are due to lateral overlap between p (or d) orbitals. A double bond between two given atoms consists of one σ and one π bond, and a triple bond is one σ and two π bonds.

Covalent bonds are also affected by the electronegativity of the connected atoms which determines the chemical polarity of the bond. Two atoms with equal electronegativity will make nonpolar covalent bonds such as H–H. An unequal relationship creates a polar covalent bond such as with H–Cl. However polarity also requires geometric asymmetry, or else dipoles may cancel out resulting in a non-polar molecule.

Covalent Structures

There are several types of structures for covalent substances, including individual molecules, molecular structures, macromolecular structures and giant covalent structures. Individual molecules have strong bonds that hold the atoms together, but there are negligible forces of attraction be-tween molecules. Such covalent substances are usually gases, for example, HCl, SO_2, CO_2, and CH_4. In molecular structures, there are weak forces of attraction. Such covalent substances are low-boiling-temperature liquids (such as ethanol), and low-melting-temperature solids (such as iodine and solid CO_2). Macromolecular structures have large numbers of atoms linked by covalent bonds in chains, including synthetic polymers such as polyethylene and nylon, and biopolymers such as proteins and starch. Network covalent structures (or giant covalent structures) contain large numbers of atoms linked in sheets (such as graphite), or 3-dimensional structures (such as diamond and quartz). These substances have high melting and boiling points, are frequently brit-tle, and tend to have high electrical resistivity. Elements that have high electronegativity, and the ability to form three or four electron pair bonds, often form such large macromolecular structures.

One- and Three-electron Bonds

Bonds with one or three electrons can be found in radical species, which have an odd number of electrons. The simplest example of a 1-electron bond is found in the dihydrogen cation, H_2^+. One-electron bonds often have about half the bond energy of a 2-electron bond, and are therefore called "half bonds". However, there are exceptions: in the case of dilithium, the bond is actually stronger for the 1-electron L_2^+ than for the 2-electron Li_2. This exception can be explained in terms of hybridization and inner-shell effects.

2e bond (e.g., CH_4)

3e bond (e.g., NO)

Comparison of the electronic structure of the three-electron bond to the conventional covalent bond.

The simplest example of three-electron bonding can be found in the helium dimer cation, He_2^+. It is considered a "half bond" because it consists of only one shared electron (rather than two); in mo-lecular orbital terms, the third electron is in an anti-bonding orbital which cancels out half of the bond formed by the other two electrons. Another example of a molecule containing a 3-electron bond, in addition to two 2-electron bonds, is nitric oxide, NO. The oxygen molecule, O_2 can also be regarded as having two 3-electron bonds and one 2-electron bond, which accounts for its para-magnetism and its formal bond order of 2. Chlorine dioxide and its heavier analogues bromine dioxide and iodine dioxide also contain three-electron bonds.

Molecules with odd-electron bonds are usually highly reactive. These types of bond are only stable between atoms with similar electronegativities.

Resonance

There are situations whereby a single Lewis structure is insufficient to explain the electron con-figuration in a molecule, hence a superposition of structures are needed. The same two atoms in such molecules can be bonded differently in different structures (a single bond in one, a double bond in another, or even none at all), resulting in a non-integer bond order. The nitrate ion is one such example with three equivalent structures. The bond between the nitrogen and each oxygen is a double bond in one structure and a single bond in the other two, so that the average bond order for each N–O interaction is $\dfrac{2 + 1 + 1}{3} = \dfrac{4}{3}$.

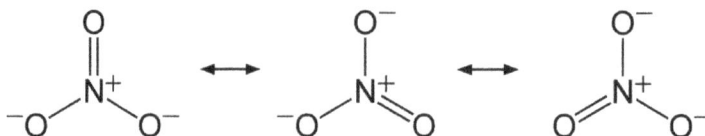

Aromaticity

In organic chemistry, when a molecule with a planar ring obeys Hückel's rule, where the number of π electrons fit the formula $4n + 2$ (where n is an integer), it attains extra stability and symmetry. In benzene, the prototypical aromatic compound, there are 6 π bonding electrons ($n = 1$, $4n + 2 = 6$). These occupy three delocalized π molecular orbitals (molecular orbital theory) or form conjugate π bonds in two resonance structures that linearly combine (valence bond theory), creating a regular hexagon exhibiting a greater stabilization than the hypothetical 1,3,5-cyclohexatriene.

In the case of heterocyclic aromatics and substituted benzenes, the electronegativity differences between different parts of the ring may dominate the chemical behaviour of aromatic ring bonds, which otherwise are equivalent.

Hypervalence

Certain molecules such as xenon difluoride and sulfur hexafluoride have higher co-ordination numbers than would be possible due to strictly covalent bonding according to the octet rule. This is explained by the three-center four-electron bond ("3c–4e") model which interprets the molecular wavefunction in terms of non-bonding highest occupied molecular orbitals in molecular orbital theory and ionic-covalent resonance in valence bond theory.

Electron-deficiency

In three-center two-electron bonds ("3c–2e") three atoms share two electrons in bonding. This type of bonding occurs in electron deficient compounds like diborane. Each such bond (2 per molecule in diborane) contains a pair of electrons which connect the boron atoms to each other in a banana shape, with a proton (nucleus of a hydrogen atom) in the middle of the bond, sharing electrons with both boron atoms. In certain cluster compounds, so-called four-center two-electron bonds also have been postulated.

Quantum Mechanical Description

After the development of quantum mechanics, two basic theories were proposed to provide a quantum description of chemical bonding: valence bond (VB) theory and molecular orbital (MO) theory. A more recent quantum description is given in terms of atomic contributions to the electronic density of states.

Covalency from Atomic Contribution to the Electronic Density of States

In COOP, COHP and BCOOP, evaluation of bond covalency is dependent on the basis set. To overcome this issue, an alternative formulation of the bond covalency can be provided in this way.

The center mass $cm(n,l,m_l,m_s)$ of an atomic orbital $|n,l,m_l,m_s\rangle$, with quantum numbers n, l, m_l, m_s, for atom A is defined as

$$cm^A(n,l,m_1,m_s) = \frac{\int_{E_0}^{E_1} E\, g_{|n,l,m_1,m_s\rangle}^A(E)\,dE}{\int_{E_0}^{E_1} g_{|n,l,m_1,m_s\rangle}^A(E)\,dE}$$

where $g_{|n,l,m_1,m_s\rangle}^A(E)$ is the contribution of the atomic orbital $|n,l,m_l,m_s\rangle$ of the atom A to the total electronic density of states $g(E)$ of the solid

$$g(E) = \sum_A \sum_{n,l} \sum_{m_1,m_s} g_{|n,l,m_1,m_s\rangle}^A(E)$$

where the outer sum runs over all atoms A of the unit cell. The energy window $[E_0, E_1]$ is chosen in such a way that it encompasses all relevant bands participating in the bond. If the range to select is unclear, it can be identified in practice by examining the molecular orbitals that describe the electron density along the considered bond.

The relative position $C_{n_A l_A, n_B l_B}$ of the center mass of $|n_A, l_A\rangle$ levels of atom A with respect to the center mass of $|n_B, l_B\rangle$ levels of atom B is given as

$$C_{n_A l_A, n_B l_B} = -\left| cm^A(n_A, l_A) - cm^B(n_B, l_B) \right|$$

where the contributions of the magnetic and spin quantum numbers are summed. According to this definition, the relative position of the A levels with respect to the B levels is

$$C_{A,B} = -\left| cm^A - cm^B \right|$$

where, for simplicity, we may omit the dependence from the principal quantum number n in the notation referring to $C_{n_A l_A, n_B l_B}$.

In this formalism, the greater the value of $C_{A,B}$, the higher the overlap of the selected atomic bands, and thus the electron density described by those orbitals gives a more covalent A–B bond. The quantity $C_{A,B}$ is denoted as the *covalency* of the A–B bond, which is specified in the same units of the energy E.

Covalent bond forms by sharing of electrons in the valence band. For example if two hydrogen atoms having one electron in their respective valence bands are brought close enough so that the bands overlap the two electrons having opposite spin fulfill the condition of a stable outer shell. They behave as if they belong to both atoms. This is why hydrogen exits in diatomic form in nature.

Since the electrons are tightly bound to a pair of atoms they are not free to move as in metallic bond it has poor electrical conductivity. Likewise in the case of oxygen atom having 6 electrons in its outer orbit it is necessary to share two electrons from each atom. It has therefore 2 covalent bonds. Similarly nitrogen having 5 electrons in its outer orbit needs to share 3 electrons with its neighbor to form 3 covalent bonds. The distance between atoms is determined by the number of bonds and the number of orbits in the atom. The bond between two carbon atoms consists of four covalent bonds. This is because to fulfill the stability of the outer shell (octet) four electrons must be shared. The best way it can happen is by having four neighboring atoms. Tetrahedron having atoms at its centre and four corners is the best example one can think of. In fact this the way the atoms of carbon are arranged in diamond. Sharing of electrons gives it its directional characteristic.

Van Der Waal Bond

This forms between atoms that behave as dipoles. The tendency to form dipole increases as more orbits are added to accommodate electrons in an atom. The centre of negatively charged moving electrons often does not coincide with the positively charged nucleus. This leads to the formation of a dipole. As it comes near another dipole bond formation occurs as a result of attraction between opposite poles. However it is much weaker than the three primary bonds described above. It can form even between atoms of noble gases having stable outer shell with 8 electrons. The nature of bonds in water molecules is also similar. However precise measurement of bond strengths shows it is a little stronger than van der Waal bond. In order to distinguish the two it is often termed as hydrogen bond. Its origin nevertheless lies in dipolar characteristic of water molecules. Such bonds are also found in organic molecules.

Bond Strength

Bond formation takes place when two atoms are close enough. Under equilibrium there is no net force acting on these. However if you try to pull them apart a restoring attractive force acts on it.

Likewise if you try to push them to bring the atoms closer there is a repulsive force which brings them back to the position of equilibrium. The net energy under such a condition should be the lowest and the net force should be zero.

The bond may therefore be visualized as an elastic spring. The stored energy is a function of its stiffness. This is given by equation 1. The negative term denotes the force of attraction and the positive term is the force of repulsion. The exponent m = 1 for ionic bond whereas m = 6 for weak van der Waal bond. Note that a lower exponent indicates a stronger bond.

$$\text{Bond Energy} = \text{attraction} + \text{repulsion} = U = -\frac{A}{r^m} + \frac{B}{r^n} \qquad (1)$$

To find out the position of equilibrium equate the first differential of U with respect r to zero. On substitution of the same in equation 1 gives the expression for bond energy. These are given by the following equations.

$$\text{For equilibrium}: \frac{dU}{dr} = \frac{Am}{r^{m+1}} - \frac{Bn}{r^{n+1}} = 0 \text{ or, } r_0 = \left(\frac{Bn}{Am}\right)^{n-m} \qquad (2)$$

On substituting (2) in (1) $U_{min} = A\left(\dfrac{m}{n}-1\right)\left[\dfrac{An}{Bm}\right]^{\left(\frac{m}{n-m}\right)}$ (3)

The strength and the stiffness of solid are primarily determined by its atomic bond. The atoms in solid are arranged in a periodic fashion. The bond energy in the neighborhood of the mean position of an atom could be represented by a Taylor series. At equilibrium the first differential of U is zero. The change in energy due to an infinitesimal displacement of the atoms by x is approximately equal to the third term of the Taylor series (higher order terms can be neglected since x is small).

$$F = Sx$$

$$\text{Stress} = \sigma = \frac{F}{r_0^2} = \frac{Sx}{r_0^2} = \frac{S}{r_0}\varepsilon$$

$$\text{Elastic Modulus} = \frac{S}{r_0}$$

Above equation shows the derivation of the relation between stiffness S and elastic modulus. Note that the force is proportional to the displacement. The constant of proportionality is the stiffness of the bond. Displacement x over r_0 is the elastic strain.

$$U(r_0 + x) = U(r_0) + \left[\frac{dU}{dr}\right]_{r=r_0} x + \frac{1}{2}\left[\frac{d^2U}{dr^2}\right]_{r=r_0} x^2 + \ldots\ldots \tag{4}$$

$$U(r_0 + x) - U(r_0) = \frac{1}{2}\left[\frac{d^2U}{dr^2}\right]_{r=r_0} x^2$$

$$\text{Force} = F = \frac{d\Delta U}{dx} = \left[\frac{d^2U}{dr^2}\right]_{r=r_0} x = Sx$$

$$\text{Stiffness} = S = \left[\frac{d^2U}{dr^2}\right] \text{ at } r = r_0 \tag{5}$$

Using the expressions for U it is possible to find an equation for S in terms of the parameters A, B, m, & n

$$\frac{dU}{dr} = \frac{Am}{r^{m+1}} - \frac{Bn}{r^{n+1}} \tag{6}$$

$$\frac{d^2U}{dr^2} = -\frac{(m+1)Am}{r^{m+2}} + \frac{(n+1)Bm}{r^{n+2}} \tag{7}$$

$$r_0 = \left(\frac{Bn}{Am}\right)^{\left(\frac{1}{n-m}\right)} \tag{8}$$

$$S = \frac{d^2U}{dr^2} = Bn(n-m)\left(\frac{Am}{Bn}\right)^{\left(\frac{n+2}{n-m}\right)} \tag{9}$$

Thus the elastic modulus is given by

$$E = S/r_0 = Bn(n-m)\left(\frac{Am}{Bn}\right)^{\left(\frac{n+2}{n-m}\right)}\left(\frac{Bn}{Am}\right)^{\left(\frac{1}{n-m}\right)} = Bn(n-m)\left(\frac{Am}{Bn}\right)^{\left(\frac{n+1}{n-m}\right)} \qquad (10)$$

Table: Strengths of four major bonds found in common solids

Bond	Example	S, N/M	E, GPa
Covalent	C-C	180	1000
Metallic	Al, Fe, Cu	75	300
Ionic	Al_2O_3	24	96
Van der Waals	$(C_2H_4)_n$	3	12
	Wax	1	4

The above table gives an idea of the strengths of various types of bonds found in different solids.

Note that carbon-carbon bond in diamond is the strongest amongst all solids. Strength of most polymeric material is derived from van der Waal bonds between chains. It is the weakest of all the four. However strength of linear polymers along the chain could be substantially high because of covalent C-C bond. In most materials except metals bonds may have mixed character. They might be partly ionic / covalent / van der Waal.

Thermal Conductivity

Thermal conductivity (often denoted k, λ, or κ) is the property of a material to conduct heat. It is evaluated primarily in terms of Fourier's Law for heat conduction.

Heat transfer occurs at a lower rate across materials of low thermal conductivity than across materials of high thermal conductivity. Correspondingly, materials of high thermal conductivity are widely used in heat sink applications and materials of low thermal conductivity are used as thermal insulation. The thermal conductivity of a material may depend on temperature. The reciprocal of thermal conductivity is called thermal resistivity.

Thermal conductivity is actually a tensor, which means it is possible to have different values in different directions.

Units of Thermal Conductivity

In SI units, thermal conductivity is measured in watts per meter-kelvin (W/(m·K)). The dimension of thermal conductivity is $M^1L^1T^{-3}\Theta^{-1}$. These variables are mass (M), length (L), time (T), and temperature (Θ). In Imperial units, thermal conductivity is measured in BTU/(hr·ft·°F).

Other units which are closely related to the thermal conductivity are in common use in the construction and textile industries. The construction industry makes use of units such as the R-value (resistance) and the U-value (transmittance). Although related to the thermal conductivity of a material used in an insulation product, R- and U-values are dependent on the thickness of the product.

Likewise the textile industry has several units including the tog and the clo which express thermal resistance of a material in a way analogous to the R-values used in the construction industry.

Measurement

There are a number of ways to measure thermal conductivity. Each of these is suitable for a limited range of materials, depending on the thermal properties and the medium temperature. There is a distinction between steady-state and transient techniques.

In general, steady-state techniques are useful when the temperature of the material does not change with time. This makes the signal analysis straightforward (steady state implies constant signals). The disadvantage is that a well-engineered experimental setup is usually needed. The Divided Bar (various types) is the most common device used for consolidated rock solids.

Experimental Values

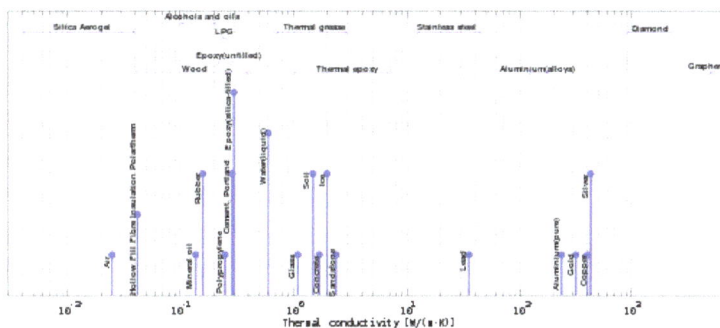

Experimental values of thermal conductivity.

Thermal conductivity is important in material science, research, electronics, building insulation and related fields, especially where high operating temperatures are achieved. Several materials are shown in the list of thermal conductivities. These should be considered approximate due to the uncertainties related to material definitions.

High energy generation rates within electronics or turbines require the use of materials with high thermal conductivity such as copper ,aluminium, and silver. On the other hand, materials with low thermal conductance, such as polystyrene and alumina, are used in building construction or in furnaces in an effort to slow the flow of heat, i.e. for insulation purposes.

Definitions

The reciprocal of thermal conductivity is *thermal resistivity*, usually expressed in kelvin-meters per watt (K·m·W^{-1}). For a given thickness of a material, that particular construction's *thermal resistance* and the reciprocal property, *thermal conductance*, can be calculated. Unfortunately, there are differing definitions for these terms.

Thermal conductivity, k, often depends on temperature. Therefore, the definitions listed below make sense when the thermal conductivity is temperature independent. Otherwise an representative mean value has to be considered.

Conductance

For general scientific use, *thermal conductance* is the quantity of heat that passes in unit time through a plate of *particular area and thickness* when its opposite faces differ in temperature by one kelvin. For a plate of thermal conductivity k, area A and thickness L, the conductance calculated is kA/L, measured in $W \cdot K^{-1}$ (equivalent to: $W/°C$). ASTM C168-15, however, defines thermal conductance as "time rate of steady state heat flow through a unit area of a material or construction induced by a unit temperature difference between the body surfaces" and defines the units as W/m^2K (Btu/(h ft^{2}F)).

The thermal conductance of that particular construction is the inverse of the thermal resistance. Thermal conductivity and conductance are analogous to electrical conductivity ($A \cdot m^{-1} \cdot V^{-1}$) and electrical conductance ($A \cdot V^{-1}$).

There is also a measure known as heat transfer coefficient: the quantity of heat that passes in unit time through a *unit area* of a plate of particular thickness when its opposite faces differ in temperature by one kelvin. The reciprocal is *thermal insulance*. In summary:

- thermal conductance = kA/L, measured in $W \cdot K^{-1}$ or in ASTM C168-15 as $W/(m^2K)$
  - thermal resistance = $L/(kA)$, measured in $K \cdot W^{-1}$ (equivalent to: $°C/W$)
- heat transfer coefficient = k/L, measured in $W \cdot K^{-1} \cdot m^{-2}$
  - thermal insulance = L/k, measured in $K \cdot m^2 \cdot W^{-1}$

The heat transfer coefficient is also known as *thermal admittance* in the sense that the material may be seen as admitting heat to flow.

Resistance

Thermal resistance is the ability of a material to resist the flow of heat.

Thermal resistance is the reciprocal of thermal conductance, i.e., lowering its value will raise the heat conduction and vice versa.

When thermal resistances occur in series, they are *additive*. Thus, when heat flows consecutively through two components each with a resistance of 3 °C/W, the total resistance is 3+3=6 °C/W.

A common engineering design problem involves the selection of an appropriate sized heat sink for a given heat source. Working in units of thermal resistance greatly simplifies the design calculation. The following formula can be used to estimate the performance:

$$R_{hs} = \frac{\Delta T}{P_{th}} - R_s$$

where:

R_{hs} is the maximum thermal resistance of the heat sink to ambient, in °C/W (equivalent to K/W)

ΔT is the required temperature difference (temperature drop), in °C

P_{th} is the thermal power (heat flow), in watts

R_s is the thermal resistance of the heat source, in °C/W

For example, if a component produces 100 W of heat, and has a thermal resistance of 0.5 °C/W, what is the maximum thermal resistance of the heat sink? Suppose the maximum temperature is 125 °C, and the ambient temperature is 25 °C; then ΔT is 100 °C. The heat sink's thermal resistance to ambient must then be 0.5 °C/W or less (total resistance component and heat sink is then 1.0 °C/W).

Transmittance

A third term, *thermal transmittance*, quantifies the thermal conductance of a structure along with heat transfer due to convection and radiation. It is measured in the same units as thermal conductance and is sometimes known as the *composite thermal conductance*. The term *U-value* is often used.

Admittance

The thermal admittance of a material, such as a building fabric, is a measure of the ability of a material to transfer heat in the presence of a temperature difference on opposite sides of the material. Thermal admittance is measured in the same units as a heat transfer coefficient, power (watts) per unit area (square meters) per temperature change (kelvins). Thermal admittance of a building fabric affects a building's thermal response to variation in outside temperature.

Influencing Factors

Temperature

The effect of temperature on thermal conductivity is different for metals and nonmetals. In metals conductivity is primarily due to free electrons. Following the Wiedemann–Franz law, thermal conductivity of metals is approximately proportional to the absolute temperature (in kelvins) times electrical conductivity. In pure metals the electrical conductivity decreases with increasing temperature and thus the product of the two, the thermal conductivity, stays approximately constant. However, as temperatures approach absolute zero, the thermal conductivity decreases sharply. In alloys the change in electrical conductivity is usually smaller and thus thermal conductivity increases with temperature, often proportionally to temperature.

On the other hand, heat conductivity in nonmetals is mainly due to lattice vibrations (phonons). Except for high quality crystals at low temperatures, the phonon mean free path is not reduced significantly at higher temperatures. Thus, the thermal conductivity of nonmetals is approximately constant at high temperatures. At low temperatures well below the Debye temperature, thermal conductivity decreases, as does the heat capacity, due to carrier scattering from defects at very low temperatures.

Chemical Phase

When a material undergoes a phase change from solid to liquid or from liquid to gas the thermal conductivity may change. An example of this would be the change in thermal conductivity that occurs when ice (thermal conductivity of 2.18 W/(m·K) at 0 °C) melts to form liquid water (thermal conductivity of 0.56 W/(m·K) at 0 °C).

Thermal Anisotropy

Some substances, such as non-cubic crystals, can exhibit different thermal conductivities along different crystal axes, due to differences in phonon coupling along a given crystal axis. Sapphire is a notable example of variable thermal conductivity based on orientation and temperature, with 35 W/(m·K) along the C-axis and 32 W/(m·K) along the A-axis. Wood generally conducts better along the grain than across it. Other examples of materials where the thermal conductivity varies with direction are metals that have undergone heavy cold pressing, laminated materials, cables, the materials used for the Space Shuttle thermal protection system, and fiber-reinforced composite structures.

When anisotropy is present, the direction of heat flow may not be exactly the same as the direction of the thermal gradient.

Electrical Conductivity

In metals, thermal conductivity approximately tracks electrical conductivity according to the Wiedemann–Franz law, as freely moving valence electrons transfer not only electric current but also heat energy. However, the general correlation between electrical and thermal conductance does not hold for other materials, due to the increased importance of phonon carriers for heat in non-metals. Highly electrically conductive silver is less thermally conductive than diamond, which is an electrical insulator, but due to its orderly array of atoms it is conductive of heat via phonons.

Magnetic Field

The influence of magnetic fields on thermal conductivity is known as the Righi-Leduc effect.

Convection

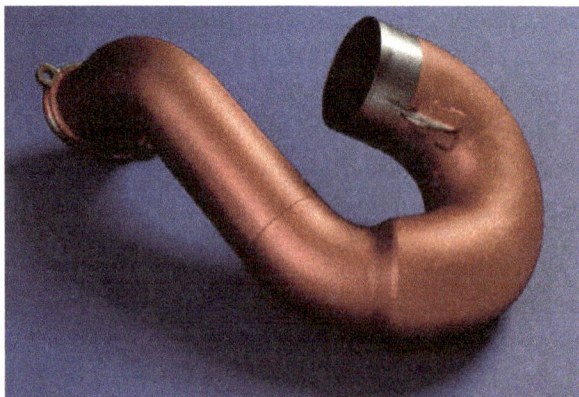

Exhaust system components with ceramic coatings having a low thermal conductivity reduce heating of nearby sensitive components

Air and other gases are generally good insulators, in the absence of convection. Therefore, many insulating materials function simply by having a large number of gas-filled pockets which prevent large-scale convection. Examples of these include expanded and extruded polystyrene (popularly referred to as "styrofoam") and silica aerogel, as well as warm clothes. Natural, biological insulators such as fur and feathers achieve similar effects by dramatically inhibiting convection of air or water near an animal's skin.

Light gases, such as hydrogen and helium typically have high thermal conductivity. Dense gases such as xenon and dichlorodifluoromethane have low thermal conductivity. An exception, sulfur hexafluoride, a dense gas, has a relatively high thermal conductivity due to its high heat capacity. Argon, a gas denser than air, is often used in insulated glazing (double paned windows) to improve their insulation characteristics.

Physical Origins

Heat flux is exceedingly difficult to control and isolate in a laboratory setting. At the atomic level, there are no simple, correct expressions for thermal conductivity. Atomically, the thermal conductivity of a system is determined by how atoms composing the system interact. There are two different approaches for calculating the thermal conductivity of a system.

- The first approach employs the Green-Kubo relations. Although this employs analytic expressions, which, in principle, can be solved, calculating the thermal conductivity of a dense fluid or solid using this relation requires the use of molecular dynamics computer simulation.

- The second approach is based on the relaxation time approach. Due to the anharmonicity within the crystal potential, the phonons in the system are known to scatter. There are three main mechanisms for scattering:

  o Boundary scattering, a phonon hitting the boundary of a system;

  o Mass defect scattering, a phonon hitting an impurity within the system and scattering;

  o Phonon-phonon scattering, a phonon breaking into two lower energy phonons or a phonon colliding with another phonon and merging into one higher-energy phonon.

Lattice Waves

Heat transport in both amorphous and crystalline dielectric solids is by way of elastic vibrations of the lattice (phonons). This transport mode is limited by the elastic scattering of acoustic phonons at lattice defects. These predictions were confirmed by the experiments of Chang and Jones on commercial glasses and glass ceramics, where the mean free paths were limited by "internal boundary scattering" to length scales of 10^{-2} cm to 10^{-3} cm.

The phonon mean free path has been associated directly with the effective relaxation length for processes without directional correlation. If V_g is the group velocity of a phonon wave packet, then the relaxation length l is defined as:

$$l = V_g t$$

where t is the characteristic relaxation time. Since longitudinal waves have a much greater phase velocity than transverse waves, V_{long} is much greater than V_{trans}, and the relaxation length or mean free path of longitudinal phonons will be much greater. Thus, thermal conductivity will be largely determined by the speed of longitudinal phonons.

Regarding the dependence of wave velocity on wavelength or frequency (dispersion), low-frequency phonons of long wavelength will be limited in relaxation length by elastic Rayleigh scattering. This type of light scattering from small particles is proportional to the fourth power of the frequency. For higher frequencies, the power of the frequency will decrease until at highest frequencies scattering is almost frequency independent. Similar arguments were subsequently generalized to many glass forming substances using Brillouin scattering.

Phonons in the acoustical branch dominate the phonon heat conduction as they have greater energy dispersion and therefore a greater distribution of phonon velocities. Additional optical modes could also be caused by the presence of internal structure (i.e., charge or mass) at a lattice point; it is implied that the group velocity of these modes is low and therefore their contribution to the lattice thermal conductivity λ_L (κ_L) is small.

Each phonon mode can be split into one longitudinal and two transverse polarization branches. By extrapolating the phenomenology of lattice points to the unit cells it is seen that the total number of degrees of freedom is 3pq when p is the number of primitive cells with q atoms/unit cell. From these only 3p are associated with the acoustic modes, the remaining 3p(q-1) are accommodated through the optical branches. This implies that structures with larger p and q contain a greater number of optical modes and a reduced λ_L.

From these ideas, it can be concluded that increasing crystal complexity, which is described by a complexity factor CF (defined as the number of atoms/primitive unit cell), decreases λ_L. Micheline Roufosse and P.G. Klemens derived the exact proportionality in their article Thermal Conductivity of Complex Dielectric Crystals at Phys. Rev. B 7, 5379–5386 (1973). This was done by assuming that the relaxation time τ decreases with increasing number of atoms in the unit cell and then scaling the parameters of the expression for thermal conductivity in high temperatures accordingly.

Describing of anharmonic effects is complicated because exact treatment as in the harmonic case is not possible and phonons are no longer exact eigensolutions to the equations of motion. Even if the state of motion of the crystal could be described with a plane wave at a particular time, its accuracy would deteriorate progressively with time. Time development would have to be described by introducing a spectrum of other phonons, which is known as the phonon decay. The two most important anharmonic effects are the thermal expansion and the phonon thermal conductivity.

Only when the phonon number ‹n› deviates from the equilibrium value ‹n›°, can a thermal current arise as stated in the following expression

$$Q_x = \frac{1}{V} \sum_{q,j} \hbar\omega \left(\langle n \rangle - \langle n \rangle^0 \right) v_x,$$

where v is the energy transport velocity of phonons. Only two mechanisms exist that can cause time variation of ‹n› in a particular region. The number of phonons that diffuse into the region from neighboring regions differs from those that diffuse out, or phonons decay inside the same region into other phonons. A special form of the Boltzmann equation

$$\frac{d\langle n\rangle}{dt} = \left(\frac{\partial\langle n\rangle}{\partial t}\right)_{diff.} + \left(\frac{\partial\langle n\rangle}{\partial t}\right)_{decay}$$

states this. When steady state conditions are assumed the total time derivate of phonon number is zero, because the temperature is constant in time and therefore the phonon number stays also constant. Time variation due to phonon decay is described with a relaxation time (τ) approximation

$$\left(\frac{\partial\langle n\rangle}{\partial t}\right)_{decay} = -\frac{\langle n\rangle - \langle n\rangle^0}{\tau},$$

which states that the more the phonon number deviates from its equilibrium value, the more its time variation increases. At steady state conditions and local thermal equilibrium are assumed we get the following equation

$$\left(\frac{\partial(n)}{\partial t}\right)_{diff.} = -v_x \frac{\partial(n)^0}{\partial T}\frac{\partial T}{\partial x}.$$

Using the relaxation time approximation for the Boltzmann equation and assuming steady-state conditions, the phonon thermal conductivity λ_L can be determined. The temperature dependence for λ_L originates from the variety of processes, whose significance for λ_L depends on the temperature range of interest. Mean free path is one factor that determines the temperature dependence for λ_L, as stated in the following equation

$$\lambda_L = \frac{1}{3V}\sum_{q,j} v(q,j)\Lambda(q,j)\frac{\partial}{\partial T}\epsilon(\omega(q,j),T),$$

where Λ is the mean free path for phonon and $\frac{\partial}{\partial T}\epsilon$ denotes the heat capacity. This equation is a result of combining the four previous equations with each other and knowing that $\langle v_x^2\rangle = \frac{1}{3}v^2$ for cubic or isotropic systems and $\Lambda = v\tau$.

At low temperatures (<10 K) the anharmonic interaction does not influence the mean free path and therefore, the thermal resistivity is determined only from processes for which q-conservation does not hold. These processes include the scattering of phonons by crystal defects, or the scattering from the surface of the crystal in case of high quality single crystal. Therefore, thermal conductance depends on the external dimensions of the crystal and the quality of the surface. Thus, temperature dependence of λ_L is determined by the specific heat and is therefore proportional to T³.

Phonon quasimomentum is defined as $\hbar q$ and differs from normal momentum because it is only defined within an arbitrary reciprocal lattice vector. At higher temperatures (10 K<T <Θ), the conservation of energy $\hbar\omega_1 = \hbar\omega_2 + \hbar\omega_3$ and quasimomentum $q_1 = q_2 + q_3 + G$, where q_1 is wave vector of the incident phonon and q_2, q_3 are wave vectors of the resultant phonons, may also involve a reciprocal lattice vector $\mathbf{G}$ complicating the energy transport process. These processes can also reverse the direction of energy transport.

Therefore, these processes are also known as Umklapp (U) processes and can only occur when phonons with sufficiently large q-vectors are excited, because unless the sum of q_2 and q_3 points outside of the Brillouin zone the momentum is conserved and the process is normal scattering (N-process). The probability of a phonon to have energy E is given by the Boltzmann distribution $P \propto e^{-E/kT}$. To U-process to occur the decaying phonon to have a wave vector q_1 that is roughly half of the diameter of the Brillouin zone, because otherwise quasimomentum would not be conserved.

Therefore, these phonons have to possess energy of $\sim k\Theta / 2$, which is a significant fraction of Debye energy that is needed to generate new phonons. The probability for this is proportional to $e^{-\Theta/bT}$, with $b = 2$. Temperature dependence of the mean free path has an exponential form $e^{\Theta/bT}$. The presence of the reciprocal lattice wave vector implies a net phonon backscattering and a resistance to phonon and thermal transport resulting finite λ_L, as it means that momentum is not conserved. Only momentum non-conserving processes can cause thermal resistance.

At high temperatures (T>Θ) the mean free path and therefore λ_L has a temperature dependence T^{-1}, to which one arrives from formula $e^{\Theta/bT}$ by making the following approximation $e^x \propto x, (x) < 1$ and writing $x = \Theta / bT$. This dependency is known as Eucken's law and originates from the temperature dependency of the probability for the U-process to occur.

Thermal conductivity is usually described by the Boltzmann equation with the relaxation time approximation in which phonon scattering is a limiting factor. Another approach is to use analytic models or molecular dynamics or Monte Carlo based methods to describe thermal conductivity in solids.

Short wavelength phonons are strongly scattered by impurity atoms if an alloyed phase is present, but mid and long wavelength phonons are less affected. Mid and long wavelength phonons carry significant fraction of heat, so to further reduce lattice thermal conductivity one has to introduce structures to scatter these phonons. This is achieved by introducing interface scattering mechanism, which requires structures whose characteristic length is longer than that of impurity atom. Some possible ways to realize these interfaces are nanocomposites and embedded nanoparticles/structures.

Electronic Thermal Conductivity

Hot electrons from higher energy states carry more thermal energy than cold electrons, while electrical conductivity is rather insensitive to the energy distribution of carriers because the amount of charge that electrons carry, does not depend on their energy. This is a physical reason for the greater sensitivity of electronic thermal conductivity to energy dependence of density of states and relaxation time, respectively.

Mahan and Sofo (*PNAS* 1996 93 (15) 7436-7439) showed that materials with a certain electron structure have reduced electron thermal conductivity. Based on their analysis one can demonstrate that if the electron density of states in the material is close to the delta-function, the electronic thermal conductivity drops to zero. By taking the following equation $\lambda_E = \lambda_0 - T\sigma S^2$, where λ_0 is the electronic thermal conductivity when the electrochemical potential gradient inside the sample is zero, as a starting point. As next step the transport coefficients are written as following:

$$\sigma = \sigma_0 I_0,$$

$$\sigma S = \left(\frac{k}{e}\right)\sigma_0 I_1$$

$$\lambda_0 = \left(\frac{k}{e}\right)^2 \sigma_0 T I_2,$$

Where $\sigma_0 = e^2 / (\hbar a_0)$ and a_0 the Bohr radius. The dimensionless integrals I_n are defined as

$$I_n = \int_{-\infty}^{\infty} \frac{e^x}{(e^x + 1)^2} s(x) x^n dx,$$

where $s(x)$ is the dimensionless transport distribution function. The integrals I_n are the moments of the function

$$P(x) = D(x)s(x), \quad D(x) = \frac{e^x}{(e^x + 1)^2},$$

where x is the energy of carriers. By substituting the previous formulas for the transport coefficient to the equation for λ_E we get the following equation

$$\lambda_E = \left(\frac{k}{e}\right)^2 \sigma_0 T \left(I_2 - \frac{I_1^2}{I_0}\right).$$

From the previous equation we see that λ_E to be zero the bracketed term containing I_n terms have to be zero. Now if we assume that

$$s(x) = f(x)\delta(x - b),$$

where δ is the Dirac delta function, I_n terms get the following expressions

$$I_0 = D(b)f(b),$$
$$I_1 = D(b)f(b)b,$$
$$I_2 = D(b)f(b)b^2.$$

By substituting these expressions to the equation for λ_E, we see that it goes to zero. Therefore, $P(x)$ has to be delta function.

Equations

In an isotropic medium the thermal conductivity is the parameter k in the Fourier expression for the heat flux

$$\vec{q} = -k\vec{\nabla}T$$

Where $\vec{q}$ is the heat flux (amount of heat flowing per second and per unit area) and $\vec{\nabla}T$ the temperature gradient. The sign in the expression is chosen so that always $k > 0$ as heat always flows from a high temperature to a low temperature. This is a direct consequence of the second law of thermodynamics.

In the one-dimensional case $q = H/A$ with H the amount of heat flowing per second through a surface with area A and the temperature gradient is dT/dx so

$$H = -kA\frac{dT}{dx}.$$

In case of a thermally-insulated bar (except at the ends) in the steady state H is constant. If A is constant as well the expression can be integrated with the result

$$HL = A\int_{T_L}^{T_H} k(T)dT$$

where T_H and T_L are the temperatures at the hot end and the cold end respectively, and L is the length of the bar. It is convenient to introduce the thermal-conductivity integral

$$I_k(T) = \int_0^T k(T')dT'.$$

The heat flow rate is then given by

$$H = \frac{A}{L}[I_k(T_H) - I_k(T_L)].$$

If the temperature difference is small k can be taken as constant. In that case

$$H = kA\frac{T_H - T_L}{L}.$$

Simple Kinetic Picture

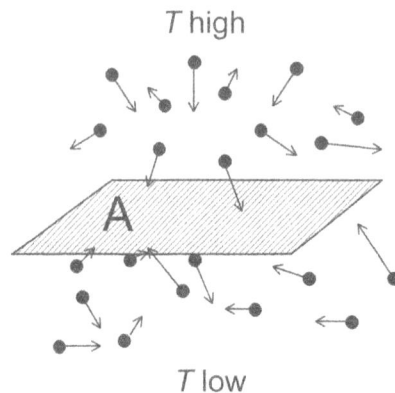

Gas atoms moving randomly through a surface.

In this section we will motivate an expression for the thermal conductivity in terms of microscopic parameters.

Consider a gas of particles of negligible volume governed hard-core interactions and within a vertical temperature gradient. The upper side is hot and the lower side cold. There is a downward energy flow because the gas atoms, going down, have a higher energy than the atoms going up. The net flow of energy per second is the heat flow H, which is proportional to the number of particles that cross the area A per second. In fact, H should also be proportional to the particle density n, the mean particle velocity v, the amount of energy transported per particle so with the heat capacity per particle c and some characteristic temperature difference ΔT. So far, in our model,

$$H \propto nvcA\Delta T.$$

The unit of H is J/s and of the right-hand side it is (particle/m³)•(m/s)•(J/(K•particle))•(m²)•(K) = J/s, so this is already of the right dimension. Only a numerical factor is missing. For ΔT we take the temperature difference of the gas between two collisions $\Delta T = l\dfrac{dT}{dz}$ where l is the mean free path.

Detailed kinetic calculations show that the numerical factor is -1/3, so, all in all,

$$H = -\frac{1}{3}nvclA\frac{dT}{dz}.$$

Comparison with the one-dimension expression for the heat flow, given above, gives an expression for the factor k

$$k = \frac{1}{3}nvcl.$$

The particle density and the heat capacity per particle can be combined as the heat capacity per unit volume

so

$$k = \frac{1}{3}vl\frac{C_V}{V_m}$$

where C_V is the molar heat capacity at constant volume and V_m the molar volume.

For the hard-core gas the mean free path is given by $l \propto \dfrac{1}{n\sigma}$ where σ is the collision cross section. So

$$k \propto \frac{c}{\sigma}v.$$

The heat capacity per particle c and the cross section σ both are temperature independent so the temperature dependence of k is determined by the T dependence of v. For a monatomic gas, with atomic mass M, v is given by $v = \sqrt{\dfrac{3RT}{M}}$. So

$$k \propto \sqrt{\frac{T}{M}}.$$

This expression also shows why gases with a low mass (hydrogen, helium) have a high thermal conductivity.

For metals at low temperatures the heat is carried mainly by the free electrons. In this case the mean velocity is the Fermi velocity which is temperature independent. The mean free path is determined by the impurities and the crystal imperfections which are temperature independent as well. So the only temperature-dependent quantity is the heat capacity c, which, in this case, is proportional to T. So

$$k = k_0 T \quad \text{(metal at low temperature)}$$

with k_0 a constant. For pure metals such as copper, silver, etc. l is large, so the thermal conductivity is high. At higher temperatures the mean free path is limited by the phonons, so the thermal conductivity tends to decrease with temperature. In alloys the density of the impurities is very high, so l and, consequently k, are small. Therefore, alloys, such as stainless steel, can be used for thermal insulation.

Metals are good conductor of heat. In fact the ratio of thermal conductivity (Kth) over electrical conductivity is constant for all metals. It is known as Wiedemann – Franz law. This states that $\frac{K_{th}}{\sigma} = 2.45 \times 10^{-8} (\text{Volt}/^{\circ}\text{K})^2$. Like electrical conductivity thermal conductivity is also inversely proportional to temperature and impurity concentration. Truly thermal conductivity should depend on specific heat, mean free path of electrons and phonons. While mean free path of electrons is of the order of 100 atomic distances, mean free path of phonon is one atomic distance. Although the contribution of electron towards specific heat is negligible it plays a crucial role in heat conduction through metals. This is because most metals have half filled valence band and overlapping empty conduction band and the number of electrons near Fermi surface is fairly large. This determines thermal conductivity.

Thermal Conductivity Measurement

There are a number of possible ways to measure thermal conductivity, each of them suitable for a limited range of materials, depending on the thermal properties and the medium temperature. Two classes of methods exist to measure the thermal conductivity of a sample: steady-state and non-steady-state (or transient) methods.

Steady-state Methods

In general, steady-state techniques perform a measurement when the temperature of the material measured does not change with time. This makes the signal analysis straightforward (steady state implies constant signals). The disadvantage is that a well-engineered experimental setup is usually needed.

In geology and geophysics, the most common method for consolidated rock samples is the divided bar. There are various modifications to these devices depending on the temperatures and pressures needed as well as sample sizes. A sample of unknown conductivity is placed between two samples of known conductivity (usually brass plates). The setup is usually vertical with the hot brass plate at the top, the sample in between then the cold brass plate at the bottom. Heat is supplied at the top and made to move downwards to stop any convection within the sample. Measurements are taken after the sample has reached to the steady state (with zero heat gradient or constant heat over entire sample), this usually takes about 30 minutes and over.

Other Steady-state Methods

For good conductors of heat, Searle's bar method can be used. For poor conductors of heat, Lees' disc method can be used.

Transient Methods

The transient techniques perform a measurement during the process of heating up. The advantage is that measurements can be made relatively quickly. Transient methods are usually carried out by needle probes.

Non-steady-state methods to measure the thermal conductivity do not require the signal to obtain a constant value. Instead, the signal is studied as a function of time. The advantage of these methods is that they can in general be performed more quickly, since there is no need to wait for a steady-state situation. The disadvantage is that the mathematical analysis of the data is in general more difficult.

Transient Plane Source Method

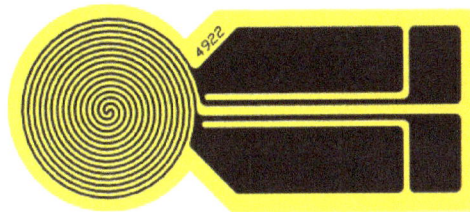

TPS sensor, model Hot Disk 4922, spiral radius about 15 mm

Transient Plane Source Method, utilizing a plane sensor and a special mathematical model describing the heat conductivity, combined with electronics, enables the method to be used to measure Thermal Transport Properties. It covers a thermal conductivity range of at least 0.01-500 W/m/K (in accordance with ISO 22007-2) and can be used for measuring various kinds of materials, such as solids, liquid, paste and thin films etc. In 2008 it was approved as an ISO-standard for measuring thermal transport properties of polymers (November 2008). This TPS standard also covers the use of this method to test both isotropic and anisotropic materials.

The Transient Plane Source technique typically employs two samples halves, in-between which the sensor is sandwiched. Normally the samples should be homogeneous, but extended use of transient plane source testing of heterogeneous material is possible, with proper selection of sensor size to maximize sample penetration. This method can also be used in a single-sided configuration, with the introduction of a known insulation material used as sensor support.

The flat sensor consists of a continuous double spiral of electrically conducting nickel (Ni) metal, etched out of a thin foil. The nickel spiral is situated between two layers of thin polyimide film Kapton. The thin Kapton films provides electrical insulation and mechanical stability to the sensor. The sensor is placed between two halves of the sample to be measured. During the measurement a constant electrical effect passes through the conducting spiral, increasing the sensor temperature. The heat generated dissipates into the sample on both sides of the sensor, at a rate depending on the thermal transport properties of the material. By recording temperature vs. time response in the sensor, the thermal conductivity, thermal diffusivity and specific heat capacity of the material can be calculated.

Modified Transient Plane Source (MTPS) Method

Modified Transient Plane Source Sensor

A variation of the above method is the Modified Transient Plane Source Method (MTPS) developed by Dr. Nancy Mathis. The device uses a one-sided, interfacial, heat reflectance sensor that applies a momentary, constant heat source to the sample. The difference between this method and traditional transient plane source technique described above is that the heating element is supported on a backing, which provides mechanical support, electrical insulation and thermal insulation. This modification provides a one-sided interfacial measurement in offering maximum flexibility in testing liquids, powders, pastes and solids.

Transient Line Source Method

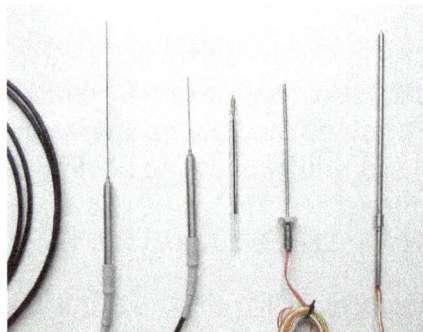

Series of needle probes used for transient line source measurements. From left to right, models TP02, TP08, a ball-point for purposes of size comparison, TP03 and TP09

The physical model behind this method is the infinite line source with constant power per unit length. The temperature profile $T(t,r)$ at a distance r at time t is as follows

$$T(t,r) = \frac{Q}{4\pi k} \text{Ei}\left(\frac{r^2}{4at}\right)$$

where

Q is the power per unit length, in [W·m⁻¹]

k is the thermal conductivity of the sample, in [W·m⁻¹·K⁻¹]

$\text{Ei}(x)$ is the exponential integral, a transcendent mathematical function

r is the radial distance to the line source

a is the thermal diffusivity, in [m²·s⁻¹]

t is the amount of time that has passed since heating has started, in [s]

When performing an experiment, one measures the temperature at a point at fixed distance, and follows that temperature in time. For large times, the exponential integral can be approximated by making use of the following relation

$$\text{Ei}(x) = -\gamma - \ln(x) + O(x^2)$$

where

$\gamma \approx 0.577$ is the Euler gamma constant

This leads to the following expression

$$T(t,r) = \frac{Q}{4\pi k}\left\{-\gamma - \ln\left(\frac{r^2}{4a}\right) + \ln(t)\right\}$$

Note that the first two terms in the brackets on the RHS are constants. Thus if the probe temperature is plotted versus the natural logarithm of time, the thermal conductivity can be determined from the slope given knowledge of Q. Typically this means ignoring the first 60 to 120 seconds of data and measuring for 600 to 1200 seconds.

Modified Transient Line Source Method

A variation on the Transient Line Source method is used for measuring the thermal conductivity of a large mass of the earth for Geothermal Heat Pump (GHP/GSHP) system design. This is generally called Ground Thermal Response Testing (TRT) by the GHP industry. Understanding the ground conductivity and thermal capacity is essential to proper GHP design, and using TRT to measure these properties was first presented in 1983 (Mogensen). The now commonly used procedure, introduced by Eklöf and Gehlin in 1996 and now approved by ASHRAE involves inserting a pipe loop deep into the ground (in a well bore, filling the anulus of the bore with a grout substance of known thermal properties, heating the fluid in the pipe loop, and measuring the temperature drop in the loop from the inlet and return pipes in the bore. The ground thermal conductivity is estimated using the line source approximation method—plotting a straight line on the log of the thermal response measured. A very stable thermal source and pumping circuit are required for this procedure.

More advanced Ground TRT methods are currently under development. The DOE is now validating a new Advanced Thermal Conductivity test said to require half the time as the existing approach, while also eliminating the requirement for a stable thermal source. This new technique is based on multi-dimensional model-based TRT data analysis.

Laser Flash Method

The laser flash method is used to measure thermal diffusivity of a thin disc in the thickness direction. This method is based upon the measurement of the temperature rise at the rear face of the thin-disc specimen produced by a short energy pulse on the front face. With a reference sample specific heat can be achieved and with known density the thermal conductivity results as follows:

$$k(T) = a(T) \cdot c_P(T) \cdot \rho(T)$$

where

k is the thermal conductivity of the sample, in $[W \cdot m^{-1} \cdot K^{-1}]$

a is the thermal diffusivity of the sample, in $[m^2 \cdot s^{-1}]$

c_P is the specific heat of the sample, in $[J \cdot kg^{-1} \cdot K^{-1}]$

ρ is the density of the sample, in $[kg \cdot m^{-3}]$

It is suitable for a multiplicity of different materials over a broad temperature range (−120 °C to 2800 °C).

3ω-method

One popular technique for electro-thermal characterization of materials is the 3ω-method, in which a thin metal structure (generally a wire or a film) is deposited on the sample to function as a resistive heater and a resistance temperature detector (RTD). The heater is driven with AC current at frequency ω, which induces periodic joule heating at frequency 2ω due to the oscillation of the AC signal during a single period. There will be some delay between the heating of the sample and the temperature response which is dependent upon the thermal properties of the sensor/sample. This temperature response is measured by logging the amplitude and phase delay of the AC voltage signal from the heater across a range of frequencies (generally accomplished using a lock-in-amplifier). Note, the phase delay of the signal is the lag between the heating signal and the temperature response. The measured voltage will contain both the fundamental and third harmonic components (ω and 3ω respectively), because the Joule heating of the metal structure induces oscillations in its resistance with frequency 2ω due to the temperature coefficient of resistance (TCR) of the metal heater/sensor as stated in the following equation:

$$V = IR = I_0 e^{i\omega t}\left(R_0 + \frac{\partial R}{\partial T}\Delta T\right) = I_0 e^{i\omega t}\left(R_0 + C_0 e^{i2\omega t}\right),$$

where C_0 is constant. Thermal conductivity is determined by the linear slope of ΔT vs. $\log(\omega)$ curve. The main advantages of the 3ω-method are minimization of radiation effects and easier acquisition of the temperature dependence of the thermal conductivity than in the steady-state techniques. Although some expertise in thin film patterning and microlithography is required, this technique is considered as the best pseudo-contact method available.

Freestanding Sensor-based 3ω-method

The freestanding sensor-based 3ω technique is proposed and developed as a candidate for the conventional 3ω method for thermophysical properties measurement. The method covers the determination of solids, powders and fluids from cryogenic temperatures to around 400 K. For solid samples, the method is applicable to both bulks and tens of micrometers thick wafers/membranes, dense or porous surfaces. The thermal conductivity and thermal effusivity can be measured using selected sensors, respectively. Two basic forms are now available: the linear source freestanding sensor and the planar source freestanding sensor. The range of thermophysical properties can be covered by different forms of the technique, with the exception that the recommended thermal conductivity range where the highest precision can be attained is 0.01 to 150 W/m•K for the linear source freestanding sensor and 500 to 8000 J/m2•K•s0.5 for the planar source freestanding sensor.

Time-domain Thermoreflectance Method

Time-domain thermoreflectance is a method by which the thermal properties of a material can be measured, most importantly thermal conductivity. This method can be applied most notably to thin film materials, which have properties that vary greatly when compared to the same materials in bulk. The idea behind this technique is that once a material is heated up, the change in the reflectance of the surface can be utilized to derive the thermal properties. The change in reflectivity is measured with respect to time, and the data received can be matched to a model which contain coefficients that correspond to thermal properties.

Measuring Devices

A thermal conductance tester, one of the instruments of gemology, determines if gems are genuine diamonds using diamond's uniquely high thermal conductivity.

Example:- Measuring Instrument of Heat Conductivity of ITP-MG4 "Zond" (Russia).

References

- Langmuir, Irving (1919-06-01). "The Arrangement of Electrons in Atoms and Molecules". Journal of the American Chemical Society. 41 (6): 868–934. doi:10.1021/ja02227a002

- Stranks, D. R.; Heffernan, M. L.; Lee Dow, K. C.; McTigue, P. T.; Withers, G. R. A. (1970). Chemistry: A structural view. Carlton, Vic.: Melbourne University Press. p. 184. ISBN 0-522-83988-6

- Chan, G. K.; Jones, R. E. (1962). "Low-Temperature Thermal Conductivity of Amorphous Solids". Physical

Review. 126 (6): 2055. Bibcode:1962PhRv..126.2055C. doi:10.1103/PhysRev.126.2055

- Lewis, Gilbert N. (1916-04-01). "The atom and the molecule". Journal of the American Chemical Society. 38 (4): 762–785. doi:10.1021/ja02261a002

- Linus Pauling, The Nature of the Chemical Bond and the Structure of Molecules and Crystals: An Introduction to Modern Structural Chemistry , Cornell University Press, 1960 ISBN 0-801-40333-2 doi:10.1021/ja01355a027

- Klemens, P.G. (1951). "The Thermal Conductivity of Dielectric Solids at Low Temperatures". Proceedings of the Royal Society of London A. 208 (1092): 108. Bibcode:1951RSPSA.208..108K. doi:10.1098/rspa.1951.0147

- Ramires, M. L. V.; Nieto de Castro, C. A.; Nagasaka, Y.; Nagashima, A.; Assael, M. J.; Wakeham, W. A. (July 6, 1994). "Standard refernce data for the thermal conductivity of water" (PDF). NIST. Retrieved 25 May 2017

- Cammarata, Antonio; Rondinelli, James M. (21 September 2014). "Covalent dependence of octahedral rotations in orthorhombic perovskite oxides". Journal of Chemical Physics. 141 (11): 114704. PMID 25240365. doi:10.1063/1.4895967

- Rowe, David Michael. Thermoelectrics handbook : macro to nano / edited by D.M. Rowe. Boca Raton: CRC/ Taylor & Francis, 2006. ISBN 0-8493-2264-2

- Pichanusakorn, P.; Bandaru, P. (2010). "Nanostructured thermoelectrics". Materials Science and Engineering: R: Reports. 67 (2–4): 19–63. doi:10.1016/j.mser.2009.10.001

- Pomeranchuk, I. (1941). "Thermal conductivity of the paramagnetic dielectrics at low temperatures". Journal of Physics (Moscow). 4: 357. ISSN 0368-3400

- Perry, R. H.; Green, D. W., eds. (1997). Perry's Chemical Engineers' Handbook (7th ed.). McGraw-Hill. Table 1–4. ISBN 978-0-07-049841-9

- IEEE Standard 98-2002 – Standard for the Preparation of Test Procedures for the Thermal Evaluation of Solid Electrical Insulating Materials, doi:10.1109/IEEESTD.2002.93617

- Qiu, L.; Li, Y. M.; Zheng, X. H.; Zhu, J.; Tang, D. W.; Wu, J. Q.; Xu, C. H. (1 December 2013). "Thermal-Conductivity Studies of Macro-porous Polymer-Derived SiOC Ceramics". International Journal of Thermophysics. 35 (1): 76–89. doi:10.1007/s10765-013-1542-8

- Chiasson, A.D. (1999). "Advances in modeling of ground source heat pump systems" (PDF). Oklahoma State University. Retrieved 2009-04-23

Thermal and Micro-structural Analysis of Metal

Thermal analysis studies changes in the properties of materials with a change in temperature. X-ray diffraction can reveal the finer details of a material as it has a smaller wavelength. Tools and techniques are an important component of any field of study. The following section elucidates the various tools and techniques that are related to physical metallurgy.

Thermal Analysis

Metals in its pure form are soft and ductile. They are rarely used in this form except in cases like electrical wires where high conductivity is of prime concern. For most applications these are subjected to various processing steps for example melting, mixing (alloying), casting, deformation and heat treatment. One of the main objectives of this course is to learn how the properties of metals and alloys can be altered by controlling such processing steps. A set of tools and techniques are necessary to monitor and evaluate the effects of process parameters on its structure and properties during each of these steps. Some of the common experimental tools and techniques which will often be referred to during this course are thermal analysis, metallography (Examination of the finer structures of metals by Optical Microscope, Scanning Electron Microscope & Transmission Electron Microscope), X-Ray Diffraction, Mechanical Properties (Tensile, Impact, Hardness, Creep & Fatigue). Therefore it would be useful to have some idea about these.

Thermal analysis is a branch of materials science where the properties of materials are studied as they change with temperature. Several methods are commonly used – these are distinguished from one another by the property which is measured:

- Dielectric thermal analysis (DEA): dielectric permittivity and loss factor
- Differential thermal analysis (DTA): temperature difference verus temperature or time
- Differential Scanning Calorimetry (DSC): heat flow changes versus temperature or time
- Dilatometry (DIL): volume changes with temperature change
- Dynamic Mechanical Analysis (DMA or DMTA) : measures storage modulus (stiffness) and loss modulus (damping) versus temperature, time and frequency
- Evolved Gas Analysis (EGA) : analysis of gases evolved during heating of a material, usually decomposition products
- Laser flash analysis (LFA): thermal diffusivity and thermal conductivity

- Thermogravimetric Analysis (TGA): mass change versus temperature or time

- Thermomechanical analysis (TMA): dimensional changes versus temperature or time

- Thermo-optical analysis (TOA): optical properties

- Derivatography: A complex method in thermal analysis

Simultaneous Thermal Analysis (STA) generally refers to the simultaneous application of Thermogravimetry (TGA) and differential scanning calorimetry (DSC) to one and the same sample in a single instrument. The test conditions are perfectly identical for the TGA and DSC signals (same atmosphere, gas flow rate, vapor pressure of the sample, heating rate, thermal contact to the sample crucible and sensor, radiation effect, etc.). The information gathered can even be enhanced by coupling the STA instrument to an Evolved Gas Analyzer (EGA) like Fourier transform infrared spectroscopy (FTIR) or mass spectrometry (MS).

Other, less common, methods measure the sound or light emission from a sample, or the electrical discharge from a dielectric material, or the mechanical relaxation in a stressed specimen. The essence of all these techniques is that the sample's response is recorded as a function of temperature (and time).

It is usual to control the temperature in a predetermined way - either by a continuous increase or decrease in temperature at a constant rate (linear heating/cooling) or by carrying out a series of determinations at different temperatures (stepwise isothermal measurements). More advanced temperature profiles have been developed which use an oscillating (usually sine or square wave) heating rate (Modulated Temperature Thermal Analysis) or modify the heating rate in response to changes in the system's properties (Sample Controlled Thermal Analysis).

In addition to controlling the temperature of the sample, it is also important to control its environment (e.g. atmosphere). Measurements may be carried out in air or under an inert gas (e.g. nitrogen or helium). Reducing or reactive atmospheres have also been used and measurements are even carried out with the sample surrounded by water or other liquids. Inverse gas chromatography is a technique which studies the interaction of gases and vapours with a surface - measurements are often made at different temperatures so that these experiments can be considered to come under the auspices of Thermal Analysis.

Atomic force microscopy uses a fine stylus to map the topography and mechanical properties of surfaces to high spatial resolution. By controlling the temperature of the heated tip and/or the sample a form of spatially resolved thermal analysis can be carried out.

Thermal analysis is also often used as a term for the study of heat transfer through structures. Many of the basic engineering data for modelling such systems comes from measurements of heat capacity and thermal conductivity.

Polymers

Polymers represent another large area in which thermal analysis finds strong applications. Thermoplastic polymers are commonly found in everyday packaging and household items, but for the analysis of the raw materials, effects of the many additive used (including stabilisers and colours) and fine-tuning of the moulding or extrusion processing used can be achieved by using DSC. An example is oxidation induction time (OIT) by DSC which can determine the amount of oxidation stabiliser present in a thermoplastic (usually a polyolefin) polymer material. Compositional analysis is often made using TGA, which can separate fillers, polymer resin and other additives. TGA can also give an indication of thermal stability and the effects of additives such as flame retardants.

Thermal analysis of composite materials, such as carbon fibre composites or glass epoxy composites are often carried out using DMA or DMTA, which can measure the stiffness of materials by determining the modulus and damping (energy absorbing) properties of the material. Aerospace companies often employ these analysers in routine quality control to ensure that products being manufactured meet the required strength specifications. Formula 1 racing car manufacturers also have similar requirements. DSC is used to determine the curing properties of the resins used in composite materials, and can also confirm whether a resin can be cured and how much heat is evolved during that process. Application of predictive kinetics analysis can help to fine-tune manufacturing processes. Another example is that TGA can be used to measure the fibre content of composites by heating a sample to remove the resin by application of heat and then determining the mass remaining.

Metals

Production of many metals (cast iron, grey iron, ductile iron, compacted graphite iron, 3000 series aluminium alloys, copper alloys, silver, and complex steels) are aided by a production technique also referred to as thermal analysis. A sample of liquid metal is removed from the furnace or ladle and poured into a sample cup with a thermocouple embedded in it. The temperature is then monitored, and the phase diagram arrests (liquidus, eutectic, and solidus) are noted. From this information chemical composition based on the phase diagram can be calculated, or the crystalline structure of the cast sample can be estimated especially for silicon morphology in hypo-eutectic Al-Si cast alloys. Strictly speaking these measurements are *cooling curves* and a form of sample controlled thermal analysis whereby the cooling rate of the sample is dependent on the cup material (usually bonded sand) and sample volume which is normally a constant due to the use of standard sized sample cups.To detect phase evolution and corresponding characteristic temperatures,cooling curve and its first derivative curve should be considered simultaneously. Examination of cooling and derivative curves is done by using appropriate data analysis software. The process consists of plotting, smoothing and curve fitting as well as identifying the reaction points and characteristic parameters. This procedure is known as Computer-Aided Cooling Curve Thermal Analysis.

Advanced techniques use differential curves to locate endothermic inflection points such as gas holes, and shrinkage, or exothermic phases such as carbides, beta crystals, inter crystalline copper, magnesium silicide, iron phosphide's and other phases as they solidify. Detection limits seem to be around 0.01% to 0.03% of volume.

In addition, integration of the area between the zero curve and the first derivative is a measure of the specific heat of that part of the solidification which can lead to rough estimates of the percent volume of a phase. (Something has to be either known or assumed about the specific heat of the phase versus the overall specific heat.) In spite of this limitation, this method is better than estimates from two dimensional micro analysis, and a lot faster than chemical dissolution.

Foods

Most foods are subjected to variations in their temperature during production, transport, storage, preparation and consumption, e.g., pasteurization, sterilization, evaporation, cooking, freezing, chilling, etc. Temperature changes cause alterations in the physical and chemical properties of food components which influence the overall properties of the final product, e.g., taste, appearance, texture and stability. Chemical reactions such as hydrolysis, oxidation or reduction may be promoted, or physical changes, such as evaporation, melting, crystallization, aggregation or gelation may occur. A better understanding of the influence of temperature on the properties of foods enables food manufacturers to optimize processing conditions and improve product quality. It is therefore important for food scientists to have analytical techniques to monitor the changes that occur in foods when their temperature varies. These techniques are often grouped under the general heading of thermal analysis. In principle, most analytical techniques can be used, or easily adapted, to monitor the temperature-dependent properties of foods, e.g., spectroscopic (NMR, UV-visible, IR spectroscopy, fluorescence), scattering (light, X-rays, neutrons), physical (mass, density, rheology, heat capacity) etc. Nevertheless, at present the term thermal analysis is usually reserved for a narrow range of techniques that measure changes in the physical properties of foods with temperature (TG/DTG, DTA,DSC and Transition temperature).

Printed Circuit Boards

Power dissipation is an important issue in present-day PCB design. Power dissipation will result in temperature difference and pose a thermal problem to a chip. In addition to the issue of reliability, excess heat will also negatively affect electrical performance and safety. The working temperature of an IC should therefore be kept below the maximum allowable limit of the worst case. In general, the temperatures of junction and ambient are 125 °C and 55 °C, respectively. The ever-shrinking chip size causes the heat to concentrate within a small area and leads to high power density. Furthermore, denser transistors gathering in a monolithic chip and higher operating frequency cause a worsening of the power dissipation. Removing the heat effectively becomes the critical issue to be resolved.

The way a metal is heated and cooled can have a profound effect on its structure and properties. In order to monitor its temperature we need a sensor. One of the most common temperature sensors consist of a pair of two different metals joined at one end. This is known as a thermocouple. If this end is heated it develops a voltage across the other ends of the wires. This can be sensed by a milliVolt (mV) meter. This depends on the temperature difference between the hot and cold ends. Two very commonly used thermocouples are (i) Pt- Pt13% Rh popularly known as R type thermocouple. This is made of a pure Pt wire joined to another wire made of Pt with 13% Rh. This can be used up to a temperature of 1600C. (ii) Chromel – Alumel popularly known as K type thermocouple. The wires here are made of two different alloys. This can be used till 1350C.

Using a pair of such a thermocouple the process of melting and solidification of a metal can easily be monitored. This shows metal kept in a crucible being heated in a furnace. The thermocouple monitors the temperature. If this is plotted against the temperature one gets a time temperature plot with a step at its melting point. During heating as long as the metal is partly liquid and partly solid the temperature of the bath does not go up. Once it is totally transformed into liquid the temperature begins to rise again. The slope of the plot on either side of the step is a function of the heat input and the specific heat of the system consisting of the container and the surrounding.

After the metal is totally molten let the furnace be switched off and the melt be allowed to cool. During this stage if the temperature is monitored and plotted against time one gets the cooling curve. This too has a step where the cooling rate (or the slope) is zero. This continues as long as the metal is partly liquid and partly solid. Once the process of solidification is complete the temperature continues to drop. Such plots tell a lot about the kinetics of transformation that goes on during solidification. This helps us understand the evolution of structure is metals during solidification and estimate its thermo physical properties.

Such nice plots could be obtained from such a simple setup if the heat effect of transformation or the amount of metal is large. However now there are sophisticated computer controlled thermal analyzers where the amount of material needed to estimate the transformation temperature and thermo physical properties is very small. One such variety is known as Differential Thermal Analyzer (DTA). This monitors the temperature of the sample of interest along with that of a standard (or a reference) material which is not likely to undergo any transformation as it is being heated or cooled. The data are stored in computer. Once transformation sets in this keeps increasing and it comes back to zero once the process is complete. The area under the peak is a function of enthalpy of transformation, specific heat, amount of material, container & surroundings. Depending on whether the process is endothermic or exothermic the peak would appear either below or above the base line.

There is another variety of thermal analyzer known as Differential Scanning Calorimeter (DSC). This has independent power supply for the sample and the reference material. The system has an intelligent controller that keeps changing the rates of heat input to both R & S so that there is no difference between the temperature of the sample under investigation and the reference material. All data are stored in computer. These are used to generate a plot of the difference in heat input rate at any instant (dH/dt) as a function of temperature during either heating or cooling cycle. Here too as long as there is no transformation dH/dt = 0. Whenever reaction or phase transformation sets in a peak appears on the plot. Depending on whether the reaction is endothermic or exothermic the peaks appear on the opposite sides of the base line.

Metals being opaque if we want to get a magnified view we would have to use a microscope which could examine its surface under reflection mode. The surface of the metal has to be carefully polished to mirror finish. This would ensure that a significant part of the incident beam gets reflected and enter the objective of the microscope. Sample preparation needs a set accessories and consumables for cutting, mounting, grinding and polishing. The quality of microstructure also depends to a great extent on the skill of the person preparing the sample. A polished surface of a pure metal may not show any feature apart from the presence of non-metallic inclusions. We need to etch this with solutions called etchant. For example steel samples are etched with 2% Nital (a very dilute solution of nitric acid in alcohol). This dissolves metals at grain boundaries to a greater extent that the grain. Sometimes the extent of attack by etchant depends on the orientations of the grains. In such cases different grains appear to have different lustre under the microscope. This is commonly known as oriented grain lustre.

The ability of a microscope to reveal fine structural details in a microstructure is defined in terms of its resolving power. This represents ability to clearly display two very closely spaced points. Often this is expressed as $\delta = \dfrac{\lambda}{2NA} = \dfrac{\lambda}{2\mu \operatorname{Sin} \alpha}$.

X-Ray Diffraction

In order to examine the finer structural details the use of smaller wavelengths of light is necessary. This could certainly be far less than the wavelength of visible light. Discovery of X-Rays having wavelengths of the order of 10nm provided a powerful tool to examine the structures of metals. X-Rays like light are electromagnetic waves with high penetrating power. It can pass through most solids. Therefore it can be used to give an idea about the internal defects in solids like castings and welds. This technique is known as radiography. The size of internal defects it can detect is of the order of 1mm. However using this to record diffraction pattern can reveal much more about the atomic arrangements in solids. Figure below gives a schematic diagram of an X-Ray source. When high voltage (~kV) is applied across a pair of electrodes in an evacuated tube electrons emanating from the cathode gets accelerated to a very high speed before hitting the anode. Usually the cathode is made of tungsten filament and a metal target is used as the anode. When the fast moving electrons decelerate rapidly it emits electromagnetic radiation. A part of this has very small wavelengths. This is called X-Rays. The wavelength of X-Rays is inversely proportional to the voltage. The derivation given in the illustration gives the wavelength of the smallest wave emanating from the source. It is called the short wavelength limit.

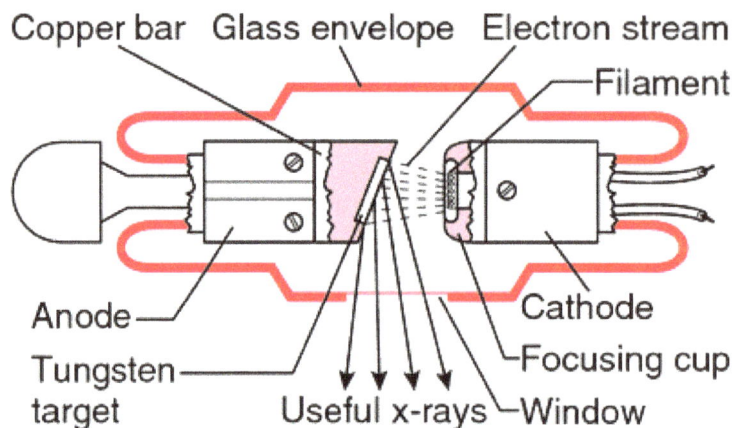

A schematic diagram of an X-Ray tube.

The rays coming out of the tube is made of radiations of different wavelengths. Typical intensity distributions of X-rays as function of wave length are given in figure below. One of these has a few peaks superimposed over a continuous spectrum. These are the characteristic peaks of the target metal.

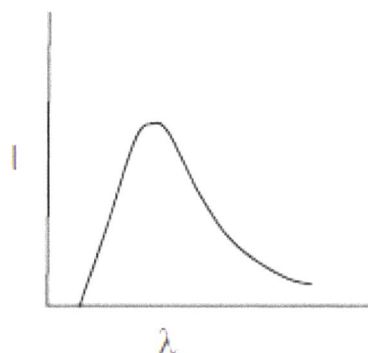

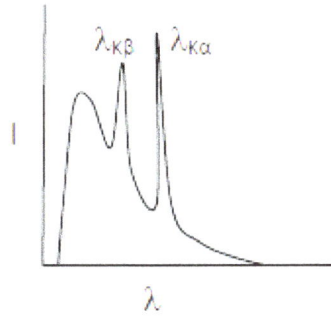

a) A continuous X-Ray spectrum giving intensity I as a function of wavelength. This is used mostly for radiography or to record diffraction pattern of single crystal. b) A characteristic X-Ray spectrum where there are superimposed sharp peaks. This occurs when kinetic energy of the electron is high enough to knock an electron out of K shell. The relation between the critical voltage and the λ_K absorption edge is similar to $\lambda_K = hc / eV_k$. When the vacant K orbit is filled up by transition of an electron from L orbit radiation is called $\lambda_{K\alpha}$. If this occurs due to transition of electron from M to K shell the radiation is called $\lambda_{K\beta}$.

Metals are crystalline. The spacing between atoms is less than1nm. Since X-Rays have wavelength of the same order metal crystal can function as diffraction gratings. The condition for diffraction to occur is given by Bragg's law described below where n is an integer:

$$2d \sin \theta = n\lambda \qquad\qquad (1)$$

Inter planar spacing (d) for a cubic crystal is a function of its lattice parameter and Miller (hkl) indices of the plane. On substitution of this in equation 1 one gets the following expression:

$$\sin^2 \theta = \frac{\lambda^2}{4a^2}((nh)^2 + (nk)^2 + (nl)^2) \qquad\qquad (2)$$

Since n too is an integer reflections may be assumed to be taking place from planes having indices <nh nk nl>. For simple cubic crystal these are 100, 110, 111, 200 etc. Note that 200 is a fictitious plane having no atoms on it. In case of a BCC crystal 200 plane has atoms. This too will take part in reflection. This would result in destructive interference. Therefore reflections from 100 plane is absent in diffraction patterns of BCC crystal. The diffraction pattern is thus a function of crystal structure.

The location and the intensity of diffracted beams depend not only on the crystal structure but also on a number other factors. These are crystal / grain size, the way the grains are oriented, the presence of residual stresses and crystal defects. This is why XRD is such a powerful tool for investigating sub-microscopic structures of solids.

Optical Microscope

The optical microscope, often referred to as light microscope, is a type of microscope which uses visible light and a system of lenses to magnify images of small samples. Optical microscopes are the oldest design of microscope and were possibly invented in their present compound form in the 17th century. Basic optical microscopes can be very simple, although there are many complex designs which aim to improve resolution and sample contrast.

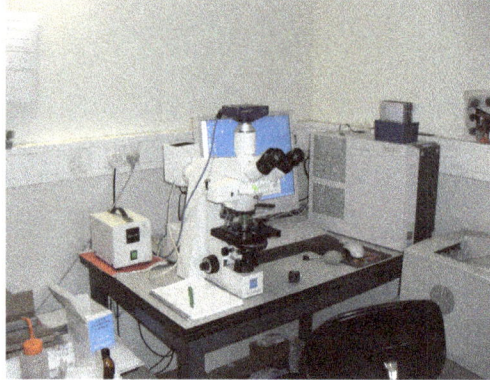

A modern microscope with a mercury bulb for fluorescence microscopy. The microscope has a digital camera, and is attached to a computer.

The image from an optical microscope can be captured by normal light-sensitive cameras to generate a micrograph. Originally images were captured by photographic film but modern developments in CMOS and charge-coupled device (CCD) cameras allow the capture of digital images. Purely digital microscopes are now available which use a CCD camera to examine a sample, showing the resulting image directly on a computer screen without the need for eyepieces.

Alternatives to optical microscopy which do not use visible light include scanning electron microscopy and transmission electron microscopy.

On 8 October 2014, the Nobel Prize in Chemistry was awarded to Eric Betzig, William Moerner and Stefan Hell for "the development of super-resolved fluorescence microscopy," which brings "optical microscopy into the nanodimension".

Types

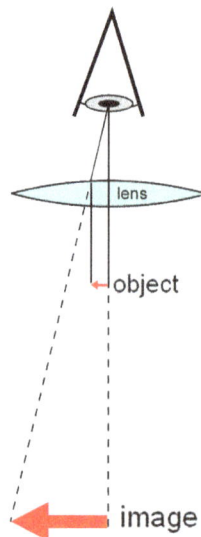

Diagram of a simple microscope

There are two basic types of optical microscopes: simple microscopes and compound microscopes. A simple microscope is one which uses a single lens for magnification, such as a magnifying glass. A compound microscope uses several lenses to enhance the magnification of an object. The vast majority of modern research microscopes are compound microscopes while some cheaper commercial digital microscopes are simple single lens microscopes. Compound microscopes can be further divided into a variety of other types of microscopes which differ in their optical configurations, cost, and intended purposes.

Simple Microscope

A simple microscope uses a lens or set of lenses to enlarge an object through angular magnification alone, giving the viewer an erect enlarged virtual image. The use of a single convex lens or groups of lenses are still found in simple magnification devices such as the magnifying glass, loupes, and eyepieces for telescopes and microscopes.

Compound Microscope

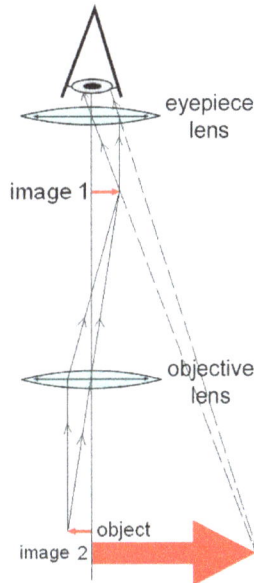

Diagram of a compound microscope

A compound microscope uses a lens close to the object being viewed to collect light (called the objective lens) which focuses a real image of the object inside the microscope (image 1). That image is then magnified by a second lens or group of lenses (called the eyepiece) that gives the viewer an enlarged inverted virtual image of the object (image 2). The use of a compound objective/eyepiece combination allows for much higher magnification. Common compound microscopes often feature exchangeable objective lenses, allowing the user to quickly adjust the magnification. A compound microscope also enables more advanced illumination setups, such as phase contrast.

Other Microscope Variants

There are many variants of the compound optical microscope design for specialized purposes. Some of these are physical design differences allowing specialization for certain purposes:

- Stereo microscope, a low-powered microscope which provides a stereoscopic view of the sample, commonly used for dissection.

- Comparison microscope, which has two separate light paths allowing direct comparison of two samples via one image in each eye.

- Inverted microscope, for studying samples from below; useful for cell cultures in liquid, or for metallography.

- Fiber optic connector inspection microscope, designed for connector end-face inspection.

Other microscope variants are designed for different illumination techniques:

- Petrographic microscope, whose design usually includes a polarizing filter, rotating stage and gypsum plate to facilitate the study of minerals or other crystalline materials whose optical properties can vary with orientation.

- Polarizing microscope, similar to the petrographic microscope.

- Phase contrast microscope, which applies the phase contrast illumination method.

- Epifluorescence microscope, designed for analysis of samples which include fluorophores.

- Confocal microscope, a widely used variant of epifluorescent illumination which uses a scanning laser to illuminate a sample for fluorescence.

- Student microscope – an often low-power microscope with simplified controls and sometimes low quality optics designed for school use or as a starter instrument for children.

- Ultramicroscope, an adapted light microscope that uses light scattering to allow viewing of tiny particles whose diameter is below or near the wavelength of visible light (around 500 nanometers); mostly obsolete since the advent of electron microscopes.

Digital Microscope

A miniature USB microscope.

A digital microscope is a microscope equipped with a digital camera allowing observation of a sample via a computer. Microscopes can also be partly or wholly computer-controlled with various levels of automation. Digital microscopy allows greater analysis of a microscope image, for example measurements of distances and areas and quantitaton of a fluorescent or histological stain.

Low-powered digital microscopes, USB microscopes, are also commercially available. These are essentially webcams with a high-powered macro lens and generally do not use transillumination. The camera attached directly to the USB port of a computer, so that the images are shown directly on the monitor. They offer modest magnifications (up to about 200×) without the need to use eyepieces, and at very low cost. High power illumination is usually provided by an LED source or sources adjacent to the camera lens.

Digital microscopy with very low light levels to avoid damage to vulnerable biological samples is available using sensitive photon-counting digital cameras. It has been demonstrated that a light source providing pairs of entangled photons may minimize the risk of damage to the most light-sensitive samples. In this application of ghost imaging to photon-sparse microscopy, the sample is illuminated with infrared photons, each of which is spatially correlated with an entangled partner in the visible band for efficient imaging by a photon-counting camera.

History

Invention

The earliest microscopes were single lens magnifying glasses with limited magnification which date at least as far back as the wide spread use of lenses in eyeglasses in the 13th century.

Compound microscopes first appeared in Europe around 1620 including one demonstrated by Cornelis Drebbel in London (around 1621) and one exhibited in Rome in 1624.

The actual inventor of the compound microscope is unknown although many claims have been made over the years. These include a claim 35 years after they appeared by Dutch spectacle-maker Johannes Zachariassen that his father, Zacharias Janssen, invented the compound microscope and/or the telescope as early as 1590. Johannes' (some claim dubious) testimony pushes the invention date so far back that Zacharias would have been a child at the time, leading to speculation that, for Johannes' claim to be true, the compound microscope would have to have been invented by Johannes' grandfather, Hans Martens. Another claim is that Janssen's competitor, Hans Lippershey (who applied for the first telescope patent in 1608) also invented the compound microscope. Other historians point to the Dutch innovator Cornelis Drebbel with his 1621 compound microscope.

Galileo Galilei is also sometimes cited as a compound microscope inventor. After 1610 he found that he could close focus his telescope to view small objects, such as flies, close up and/or could look through the wrong end in reverse to magnify small objects. The only drawback was that his 2 foot long telescope had to be extended out to 6 feet to view objects that close. After seeing the compound microscope built by Drebbel exhibited in Rome in 1624, Galileo built his own improved version. In 1625 Giovanni Faber coined the name *microscope* for the compound microscope Galileo submitted to the Accademia dei Lincei in 1624 (Galileo had called it the "*occhiolino*" or "*little eye*"). Faber coined the name from the Greek words (micron) meaning "small", and (skopein) meaning "to look at", a name meant to be analogous with "telescope", another word coined by the Linceans.

Christiaan Huygens, another Dutchman, developed a simple 2-lens ocular system in the late 17th century that was achromatically corrected, and therefore a huge step forward in microscope development. The Huygens ocular is still being produced to this day, but suffers from a small field size, and other minor disadvantages.

Popularization

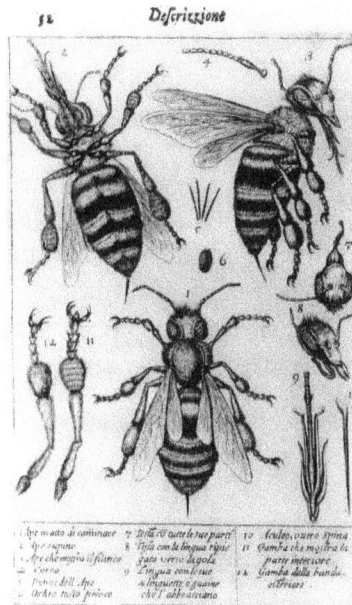

The oldest published image known to have been made with a microscope: bees by Francesco Stelluti, 1630

Antonie van Leeuwenhoek (1632–1724) is credited with bringing the microscope to the attention of biologists, even though simple magnifying lenses were already being produced in the 16th century. Van Leeuwenhoek's home-made microscopes were simple microscopes, with a single very small, yet strong lens. They were awkward in use, but enabled van Leeuwenhoek to see detailed images. It took about 150 years of optical development before the compound microscope was able to provide the same quality image as van Leeuwenhoek's simple microscopes, due to difficulties in configuring multiple lenses. In the 1850s John Leonard Riddell, Professor of Chemistry at Tulane University, invented the first practical binocular microscope while carrying out one of the earliest and most extensive American microscopic investigations of cholera.

Lighting Techniques

While basic microscope technology and optics have been available for over 400 years it is much more recently that techniques in sample illumination were developed to generate the high quality images seen today.

In August 1893 August Köhler developed Köhler illumination. This method of sample illumination gives rise to extremely even lighting and overcomes many limitations of older techniques of sample illumination. Before development of Köhler illumination the image of the light source, for example a lightbulb filament, was always visible in the image of the sample.

The Nobel Prize in physics was awarded to Dutch physicist Frits Zernike in 1953 for his development of phase contrast illumination which allows imaging of transparent samples. By using interference rather than absorption of light, extremely transparent samples, such as live mammalian cells, can be imaged without having to use staining techniques. Just two years later, in 1955, Georges Nomarski published the theory for differential interference contrast microscopy, another interference-based imaging technique.

Fluorescence Microscopy

Modern biological microscopy depends heavily on the development of fluorescent probes for specific structures within a cell. In contrast to normal transilluminated light microscopy, in fluorescence microscopy the sample is illuminated through the objective lens with a narrow set of wavelengths of light. This light interacts with fluorophores in the sample which then emit light of a longer wavelength. It is this emitted light which makes up the image.

Since the mid 20th century chemical fluorescent stains, such as DAPI which binds to DNA, have been used to label specific structures within the cell. More recent developments include immunofluorescence, which uses fluorescently labelled antibodies to recognise specific proteins within a sample, and fluorescent proteins like GFP which a live cell can express making it fluorescent.

Components

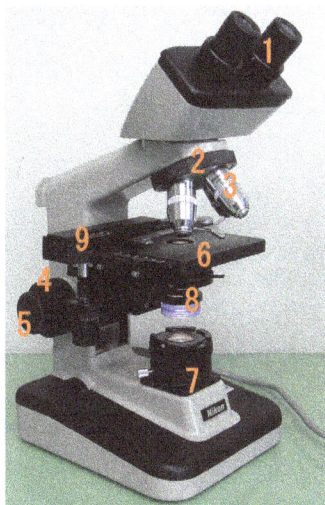

Basic optical transmission microscope elements (1990s)

All modern optical microscopes designed for viewing samples by transmitted light share the same basic components of the light path. In addition, the vast majority of microscopes have the same 'structural' components (numbered below according to the image on the right):

- Eyepiece (ocular lens) (1)

- Objective turret, revolver, or revolving nose piece (to hold multiple objective lenses) (2)

- Objective lenses (3)

- Focus knobs (to move the stage)

  o Coarse adjustment (4)

  o Fine adjustment (5)

- Stage (to hold the specimen) (6)

- Light source (a light or a mirror) (7)

- Diaphragm and condenser (8)

- Mechanical stage (9)

Eyepiece (Ocular Lens)

The eyepiece, or ocular lens, is a cylinder containing two or more lenses; its function is to bring the image into focus for the eye. The eyepiece is inserted into the top end of the body tube. Eyepieces are interchangeable and many different eyepieces can be inserted with different degrees of magnification. Typical magnification values for eyepieces include 5×, 10× (the most common), 15X and 20×. In some high performance microscopes, the optical configuration of the objective lens and eyepiece are matched to give the best possible optical performance. This occurs most commonly with apochromatic objectives.

Objective Turret (Revolver or Revolving Nose Piece)

Objective turret, revolver, or revolving nose piece is the part that holds the set of objective lenses. It allows the user to switch between objective lenses.

Objective

At the lower end of a typical compound optical microscope, there are one or more objective lenses that collect light from the sample. The objective is usually in a cylinder housing containing a glass single or multi-element compound lens. Typically there will be around three objective lenses screwed into a circular nose piece which may be rotated to select the required objective lens. These arrangements are designed to be parfocal, which means that when one changes from one lens to another on a microscope, the sample stays in focus. Microscope objectives are characterized by two parameters, namely, magnification and numerical aperture. The former typically ranges from 5× to 100× while the latter ranges from 0.14 to 0.7, corresponding to focal lengths of about 40 to 2 mm, respectively. Objective lenses with higher magnifications normally have a higher numerical aperture and a shorter depth of field in the resulting image. Some high performance objective lenses may require matched eyepieces to deliver the best optical performance.

Oil Immersion Objective

Two Leica oil immersion microscope objective lenses: 100× (left) and 40× (right)

Some microscopes make use of oil-immersion objectives or water-immersion objectives for greater resolution at high magnification. These are used with index-matching material such as immersion oil or water and a matched cover slip between the objective lens and the sample. The refractive index of the index-matching material is higher than air allowing the objective lens to have a larger numerical aperture (greater than 1) so that the light is transmitted from the specimen to the outer face of the objective lens with minimal refraction. Numerical apertures as high as 1.6 can be achieved. The larger numerical aperture allows collection of more light making detailed observation of smaller details possible. An oil immersion lens usually has a magnification of 40 to 100×.

Focus Knobs

Adjustment knobs move the stage up and down with separate adjustment for coarse and fine focusing. The same controls enable the microscope to adjust to specimens of different thickness. In older designs of microscopes, the focus adjustment wheels move the microscope tube up or down relative to the stand and had a fixed stage.

Frame

The whole of the optical assembly is traditionally attached to a rigid arm, which in turn is attached to a robust U-shaped foot to provide the necessary rigidity. The arm angle may be adjustable to allow the viewing angle to be adjusted.

The frame provides a mounting point for various microscope controls. Normally this will include controls for focusing, typically a large knurled wheel to adjust coarse focus, together with a smaller knurled wheel to control fine focus. Other features may be lamp controls and/or controls for adjusting the condenser.

Stage

The stage is a platform below the objective which supports the specimen being viewed. In the center of the stage is a hole through which light passes to illuminate the specimen. The stage usually has arms to hold slides (rectangular glass plates with typical dimensions of 25×75 mm, on which the specimen is mounted).

At magnifications higher than 100× moving a slide by hand is not practical. A mechanical stage, typical of medium and higher priced microscopes, allows tiny movements of the slide via control knobs that reposition the sample/slide as desired. If a microscope did not originally have a mechanical stage it may be possible to add one.

All stages move up and down for focus. With a mechanical stage slides move on two horizontal axes for positioning the specimen to examine specimen details.

Focusing starts at lower magnification in order to center the specimen by the user on the stage. Moving to a higher magnification requires the stage to be moved higher vertically for re-focus at the higher magnification and may also require slight horizontal specimen position adjustment. Horizontal specimen position adjustments are the reason for having a mechanical stage.

Due to the difficulty in preparing specimens and mounting them on slides, for children it's best to begin with prepared slides that are centered and focus easily regardless of the focus level used.

Light Source

Many sources of light can be used. At its simplest, daylight is directed via a mirror. Most microscopes, however, have their own adjustable and controllable light source – often a halogen lamp, although illumination using LEDs and lasers are becoming a more common provision.

Condenser

The condenser is a lens designed to focus light from the illumination source onto the sample. The condenser may also include other features, such as a diaphragm and/or filters, to manage the quality and intensity of the illumination. For illumination techniques like dark field, phase contrast and differential interference contrast microscopy additional optical components must be precisely aligned in the light path.

Magnification

The actual power or magnification of a compound optical microscope is the product of the powers of the ocular (eyepiece) and the objective lens. The maximum normal magnifications of the ocular and objective are 10× and 100× respectively, giving a final magnification of 1,000×.

Magnification and Micrographs

When using a camera to capture a micrograph the effective magnification of the image must take into account the size of the image. This is independent of whether it is on a print from a film negative or displayed digitally on a computer screen.

In the case of photographic film cameras the calculation is simple; the final magnification is the product of: the objective lens magnification, the camera optics magnification and the enlargement factor of the film print relative to the negative. A typical value of the enlargement factor is around 5× (for the case of 35mm film and a 15x10 cm (6×4 inch) print).

In the case of digital cameras the size of the pixels in the CMOS or CCD detector and the size of the pixels on the screen have to be known. The enlargement factor from the detector to the pixels on screen can then be calculated. As with a film camera the final magnification is the product of: the objective lens magnification, the camera optics magnification and the enlargement factor.

Operation

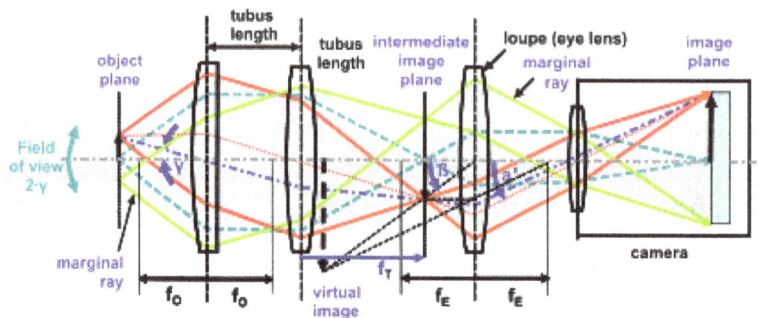

Optical path in a typical microscope

The optical components of a modern microscope are very complex and for a microscope to work well, the whole optical path has to be very accurately set up and controlled. Despite this, the basic operating principles of a microscope are quite simple.

The objective lens is, at its simplest, a very high-powered magnifying glass, i.e. a lens with a very short focal length. This is brought very close to the specimen being examined so that the light from the specimen comes to a focus about 160 mm inside the microscope tube. This creates an enlarged image of the subject. This image is inverted and can be seen by removing the eyepiece and placing a piece of tracing paper over the end of the tube. By carefully focusing a brightly lit specimen, a highly enlarged image can be seen. It is this real image that is viewed by the eyepiece lens that provides further enlargement.

In most microscopes, the eyepiece is a compound lens, with one component lens near the front and one near the back of the eyepiece tube. This forms an air-separated couplet. In many designs, the virtual image comes to a focus between the two lenses of the eyepiece, the first lens bringing the real image to a focus and the second lens enabling the eye to focus on the virtual image.

In all microscopes the image is intended to be viewed with the eyes focused at infinity (mind that the position of the eye in the figure is determined by the eye's focus). Headaches and tired eyes after using a microscope are usually signs that the eye is being forced to focus at a close distance rather than at infinity.

Illumination Techniques

Many techniques are available which modify the light path to generate an improved contrast image from a sample. Major techniques for generating increased contrast from the sample include cross-polarized light, dark field, phase contrast and differential interference contrast illumination. A recent technique (Sarfus) combines cross-polarized light and specific contrast-enhanced slides for the visualization of nanometric samples.

Four examples of transilumination techniques used to generate contrast in a sample of tissue paper. 1.559 µm/pixel.

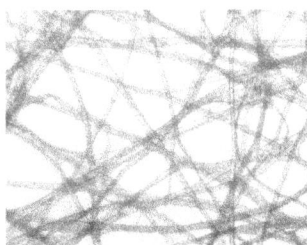

Bright field illumination, sample contrast comes from absorbance of light in the sample.

Cross-polarized light illumination, sample contrast comes from rotation of polarized light through the sample.

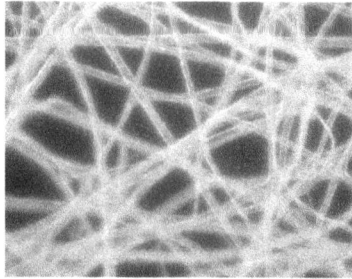

Dark field illumination, sample contrast comes from light scattered by the sample.

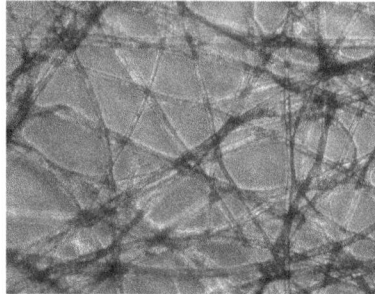

Phase contrast illumination, sample contrast comes from interference of different path lengths of light through the sample.

Other Techniques

Modern microscopes allow more than just observation of transmitted light image of a sample; there are many techniques which can be used to extract other kinds of data. Most of these require additional equipment in addition to a basic compound microscope.

- Reflected light, or incident, illumination (for analysis of surface structures)

- Fluorescence microscopy, both:

  - Epifluorescence microscopy

  - Confocal microscopy

- Microspectroscopy (where a UV-visible spectrophotometer is integrated with an optical microscope)

- Ultraviolet microscopy

- Near-Infrared microscopy

- Multiple transmission microscopy for contrast enhancement and aberration reduction.

- Automation (for automatic scanning of a large sample or image capture).

Applications

Optical microscopy is used extensively in microelectronics, nanophysics, biotechnology, pharmaceutic research, mineralogy and microbiology.

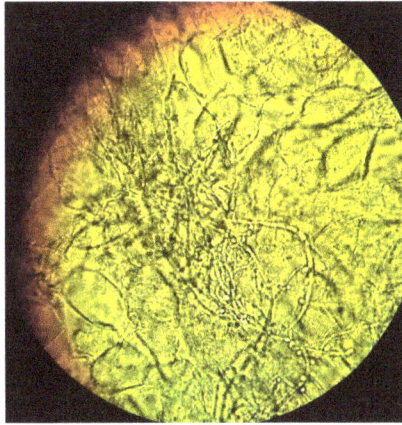

A 40x magnification image of cells in a medical smear test taken through an optical microscope using a wet mount technique, placing the specimen on a glass slide and mixing with a salt solution

Optical microscopy is used for medical diagnosis, the field being termed histopathology when dealing with tissues, or in smear tests on free cells or tissue fragments.

In industrial use, binocular microscopes are common. Aside from applications needing true depth perception, the use of dual eyepieces reduces eye strain associated with long workdays at a microscopy station. In certain applications, long-working-distance or long-focus microscopes are beneficial. An item may need to be examined behind a window, or industrial subjects may be a hazard to the objective. Such optics resemble telescopes with close-focus capabilities.

Measuring microscopes are used for precision measurement. There are two basic types. One has a reticle graduated to allow measuring distances in the focal plane. The other (and older) type has simple crosshairs and a micrometer mechanism for moving the subject relative to the microscope.

Limitations

At very high magnifications with transmitted light, point objects are seen as fuzzy discs surrounded by diffraction rings. These are called Airy disks. The *resolving power* of a microscope is taken as the ability to distinguish between two closely spaced Airy disks (or, in other words the ability of the microscope to reveal adjacent structural detail as distinct and separate). It is these impacts of diffraction that limit the ability to resolve fine details. The extent and magnitude of the diffraction patterns are affected by both the wavelength of light (λ), the refractive materials used to manufacture the objective lens and the numerical aperture (NA) of the objective lens. There is therefore a finite limit beyond which it is impossible to resolve separate points in the objective field, known as the diffraction limit. Assuming that optical aberrations in the whole optical set-up are negligible, the resolution d, can be stated as:

$$d = \frac{\lambda}{2NA}$$

Usually a wavelength of 550 nm is assumed, which corresponds to green light. With air as the external medium, the highest practical NA is 0.95, and with oil, up to 1.5. In practice the lowest value of d obtainable with conventional lenses is about 200 nm. A new type of lens using multiple scattering of light allowed to improve the resolution to below 100 nm.

Surpassing the Resolution Limit

Multiple techniques are available for reaching resolutions higher than the transmitted light limit described above. Holographic techniques, as described by Courjon and Bulabois in 1979, are also capable of breaking this resolution limit, although resolution was restricted in their experimental analysis.

Using fluorescent samples more techniques are available. Examples include Vertico SMI, near field scanning optical microscopy which uses evanescent waves, and stimulated emission depletion. In 2005, a microscope capable of detecting a single molecule was described as a teaching tool.

Despite significant progress in the last decade, techniques for surpassing the diffraction limit remain limited and specialized.

While most techniques focus on increases in lateral resolution there are also some techniques which aim to allow analysis of extremely thin samples. For example, sarfus methods place the thin sample on a contrast-enhancing surface and thereby allows to directly visualize films as thin as 0.3 nanometers.

Structured Illumination SMI

SMI (spatially modulated illumination microscopy) is a light optical process of the so-called point spread function (PSF) engineering. These are processes which modify the PSF of a microscope in a suitable manner to either increase the optical resolution, to maximize the precision of distance measurements of fluorescent objects that are small relative to the wavelength of the illuminating light, or to extract other structural parameters in the nanometer range.

Localization Microscopy SPDMphymod

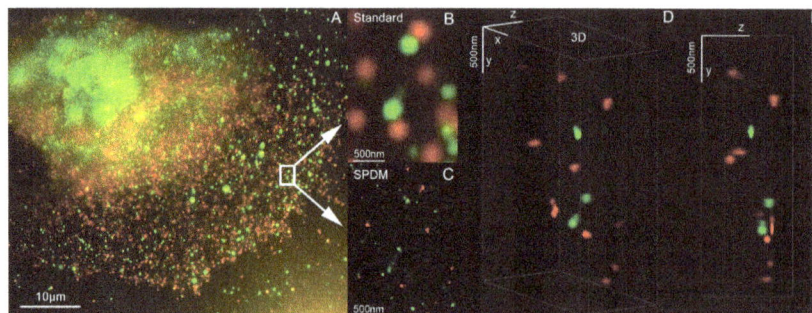

3D dual color super resolution microscopy with Her2 and Her3 in breast cells, standard dyes: Alexa 488, Alexa 568 LIMON

SPDM (spectral precision distance microscopy), the basic localization microscopy technology is a light optical process of fluorescence microscopy which allows position, distance and angle measurements on "optically isolated" particles (e.g. molecules) well below the theoretical limit of resolution for light microscopy. "Optically isolated" means that at a given point in time, only a single particle/molecule within a region of a size determined by conventional optical resolution (typically approx. 200–250 nm diameter) is being registered. This is possible when molecules within such a region all carry different spectral markers (e.g. different colors or other usable differences in the light emission of different particles).

Many standard fluorescent dyes like GFP, Alexa dyes, Atto dyes, Cy2/Cy3 and fluorescein molecules can be used for localization microscopy, provided certain photo-physical conditions are present. Using this so-called SPDMphymod (physically modifiable fluorophores) technology a single laser wavelength of suitable intensity is sufficient for nanoimaging.

3D Super Resolution Microscopy

3D super resolution microscopy with standard fluorescent dyes can be achieved by combination of localization microscopy for standard fluorescent dyes SPDMphymod and structured illumination SMI.

STED

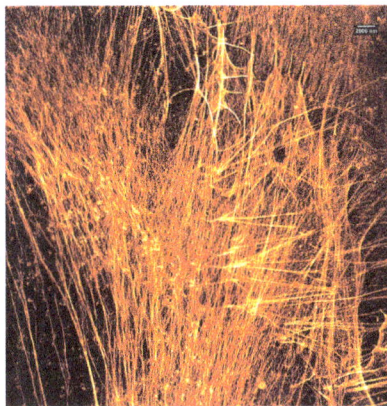

Stimulated emission depletion (STED) microscopy image of actin filaments within a cell.

Stimulated emission depletion is a simple example of how higher resolution surpassing the diffraction limit is possible, but it has major limitations. STED is a fluorescence microscopy technique which uses a combination of light pulses to induce fluorescence in a small sub-population of fluorescent molecules in a sample. Each molecule produces a diffraction-limited spot of light in the image, and the centre of each of these spots corresponds to the location of the molecule. As the number of fluorescing molecules is low the spots of light are unlikely to overlap and therefore can be placed accurately. This process is then repeated many times to generate the image. Stefan Hell of the Max Planck Institute for Biophysical Chemistry was awarded the 10th German Future Prize in 2006 and Nobel Prize for Chemistry in 2014 for his development of the STED microscope and associated methodologies.

Alternatives

In order to overcome the limitations set by the diffraction limit of visible light other microscopes have been designed which use other waves.

- Atomic force microscope (AFM)
- Scanning electron microscope (SEM)
- Scanning ion-conductance microscopy (SICM)
- Scanning tunneling microscope (STM)

- Transmission electron microscopy (TEM)

- Ultraviolet microscope

- X-ray microscope

It is important to note that higher frequency waves have limited interaction with matter, for example soft tissues are relatively transparent to X-rays resulting in distinct sources of contrast and different target applications.

The use of electrons and X-rays in place of light allows much higher resolution – the wavelength of the radiation is shorter so the diffraction limit is lower. To make the short-wavelength probe non-destructive, the atomic beam imaging system (atomic nanoscope) has been proposed and widely discussed in the literature, but it is not yet competitive with conventional imaging systems.

STM and AFM are scanning probe techniques using a small probe which is scanned over the sample surface. Resolution in these cases is limited by the size of the probe; micromachining techniques can produce probes with tip radii of 5–10 nm.

Additionally, methods such as electron or X-ray microscopy use a vacuum or partial vacuum, which limits their use for live and biological samples (with the exception of an environmental scanning electron microscope). The specimen chambers needed for all such instruments also limits sample size, and sample manipulation is more difficult. Color cannot be seen in images made by these methods, so some information is lost. They are however, essential when investigating molecular or atomic effects, such as age hardening in aluminium alloys, or the microstructure of polymers.

Transmission Electron Microscopy

Transmission electron microscopy (TEM, also sometimes conventional transmission electron microscopy or CTEM) is a microscopy technique in which a beam of electrons is transmitted through a specimen to form an image. The specimen is most often an ultrathin section less than 100 nm thick or a suspension on a grid. An image is formed from the interaction of the electrons with the sample as the beam is transmitted through the specimen. The image is then magnified and focused onto an imaging device, such as a fluorescent screen, a layer of photographic film, or a sensor such as a charge-coupled device.

Transmission electron microscopes are capable of imaging at a significantly higher resolution than light microscopes, owing to the smaller de Broglie wavelength of electrons. This enables the instrument to capture fine detail—even as small as a single column of atoms, which is thousands of times smaller than a resolvable object seen in a light microscope. Transmission electron microscopy is a major analytical method in the physical, chemical and biological sciences. TEMs find application in cancer research, virology, and materials science as well as pollution, nanotechnology and semiconductor research.

At lower magnifications TEM image contrast is due to differential absorption of electrons by the material due to differences in composition or thickness of the material. At higher magnifications complex wave interactions modulate the intensity of the image, requiring expert analysis of observed images. Alternate modes of use allow for the TEM to observe modulations in chemical identity, crystal orientation, electronic structure and sample induced electron phase shift as well as the regular absorption based imaging.

The first TEM was demonstrated by Max Knoll and Ernst Ruska in 1931, with this group developing the first TEM with resolution greater than that of light in 1933 and the first commercial TEM in 1939. In 1986, Ruska was awarded the Nobel Prize in physics for the development of transmission electron microscopy.

History

Initial Development

The first practical TEM, originally installed at IG Farben-Werke and now on display at the Deutsches Museum in Munich, Germany

In 1873, Ernst Abbe proposed that the ability to resolve detail in an object was limited approximately by the wavelength of the light used in imaging or a few hundred nanometers for visible light microscopes. Developments in ultraviolet (UV) microscopes, led by Köhler and Rohr, increased resolving power by a factor of two. However this required expensive quartz optics, due to the absorption of UV by glass. It was believed that obtaining an image with sub-micrometer information was not possible due to this wavelength constraint.

In 1858 Plücker observed the deflection of "cathode rays" (electrons) with the use of magnetic fields. This effect was utilized by Ferdinand Braun in 1897 to build simple cathode ray oscilloscopes (CROs) measuring devices. In 1891 Riecke noticed that the cathode rays could be focused by magnetic fields, allowing for simple electromagnetic lens designs. In 1926 Hans Busch published work extending this theory and showed that the lens maker's equation could, with appropriate assumptions, be applied to electrons.

In 1928, at the Technical University of Berlin, Adolf Matthias, Professor of High voltage Technology and Electrical Installations, appointed Max Knoll to lead a team of researchers to advance the CRO design. The team consisted of several PhD students including Ernst Ruska and Bodo von

Borries. The research team worked on lens design and CRO column placement, to optimize parameters to construct better CROs, and make electron optical components to generate low magnification (nearly 1:1) images. In 1931 the group successfully generated magnified images of mesh grids placed over the anode aperture. The device used two magnetic lenses to achieve higher magnifications, arguably creating the first electron microscope. In that same year, Reinhold Rudenberg, the scientific director of the Siemens company, patented an electrostatic lens electron microscope.

Improving Resolution

At the time, electrons were understood to be charged particles of matter; the wave nature of electrons was not fully realized until the publication of the De Broglie hypothesis in 1927. The research group was unaware of this publication until 1932, when they quickly realized that the De Broglie wavelength of electrons was many orders of magnitude smaller than that for light, theoretically allowing for imaging at atomic scales. In April 1932, Ruska suggested the construction of a new electron microscope for direct imaging of specimens inserted into the microscope, rather than simple mesh grids or images of apertures. With this device successful diffraction and normal imaging of an aluminium sheet was achieved. However the magnification achievable was lower than with light microscopy. Magnifications higher than those available with a light microscope were achieved in September 1933 with images of cotton fibers quickly acquired before being damaged by the electron beam.

At this time, interest in the electron microscope had increased, with other groups, such as Paul Anderson and Kenneth Fitzsimmons of Washington State University, and Albert Prebus and James Hillier at the University of Toronto, who constructed the first TEMs in North America in 1935 and 1938, respectively, continually advancing TEM design.

Research continued on the electron microscope at Siemens in 1936, where the aim of the research was the development and improvement of TEM imaging properties, particularly with regard to biological specimens. At this time electron microscopes were being fabricated for specific groups, such as the "EM1" device used at the UK National Physical Laboratory. In 1939 the first commercial electron microscope, pictured, was installed in the Physics department of IG Farben-Werke. Further work on the electron microscope was hampered by the destruction of a new laboratory constructed at Siemens by an air-raid, as well as the death of two of the researchers, Heinz Müller and Friedrick Krause during World War II.

Further Research

After World War II, Ruska resumed work at Siemens, where he continued to develop the electron microscope, producing the first microscope with 100k magnification. The fundamental structure of this microscope design, with multi-stage beam preparation optics, is still used in modern microscopes. The worldwide electron microscopy community advanced with electron microscopes being manufactured in Manchester UK, the USA (RCA), Germany (Siemens) and Japan (JEOL). The first international conference in electron microscopy was in Delft in 1949, with more than one hundred attendees. Later conferences included the "First" international conference in Paris, 1950 and then in London in 1954.

With the development of TEM, the associated technique of scanning transmission electron microscopy (STEM) was re-investigated and did not become developed until the 1970s, with Albert Crewe at the University of Chicago developing the field emission gun and adding a high quality objective lens to create the modern STEM. Using this design, Crewe demonstrated the ability to image atoms using annular dark-field imaging. Crewe and coworkers at the University of Chicago developed the cold field electron emission source and built a STEM able to visualize single heavy atoms on thin carbon substrates. In 2008, Jannick Meyer et al. described the direct visualization of light atoms such as carbon and even hydrogen using TEM and a clean single-layer graphene substrate.

Background

Electrons

Theoretically, the maximum resolution, d, that one can obtain with a light microscope has been limited by the wavelength of the photons that are being used to probe the sample, λ, and the numerical aperture of the system, NA.

$$d = \frac{\lambda}{2n \sin \alpha} \approx \frac{\lambda}{2NA}$$

where n is the index of refraction of the medium in which the lens is working and α is the maximum half-angle of the cone of light that can enter the lens. Early twentieth century scientists theorized ways of getting around the limitations of the relatively large wavelength of visible light (wavelengths of 400–700 nanometers) by using electrons. Like all matter, electrons have both wave and particle properties (as theorized by Louis-Victor de Broglie), and their wave-like properties mean that a beam of electrons can be made to behave like a beam of electromagnetic radiation. The wavelength of electrons is related to their kinetic energy via the de Broglie equation. An additional correction must be made to account for relativistic effects, as in a TEM an electron's velocity approaches the speed of light, c.

$$\lambda_e \approx \frac{h}{\sqrt{2m_0 E \left(1 + \frac{E}{2m_0 c^2} \right)}}$$

where, h is Planck's constant, m_0 is the rest mass of an electron and E is the energy of the accelerated electron. Electrons are usually generated in an electron microscope by a process known as thermionic emission from a filament, usually tungsten, in the same manner as a light bulb, or alternatively by field electron emission. The electrons are then accelerated by an electric potential (measured in volts) and focused by electrostatic and electromagnetic lenses onto the sample. The transmitted beam contains information about electron density, phase and periodicity; this beam is used to form an image.

Source Formation

From the top down, the TEM consists of an emission source, which may be a tungsten filament or needle, or a lanthanum hexaboride (LaB_6) single crystal source. The gun is connected to a high voltage source (typically ~100–300 kV) and, given sufficient current, the gun will begin to emit electrons either by thermionic or field electron emission into the vacuum. The electron source is typically mounted in a Wehnelt cylinder to provide preliminary focus of the emitted electrons into a beam. The upper lenses of the TEM then further focus the electron beam to the desired size and location.

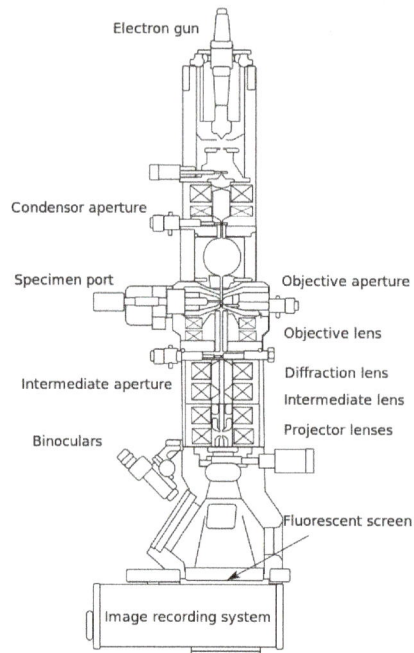

Layout of optical components in a basic TEM

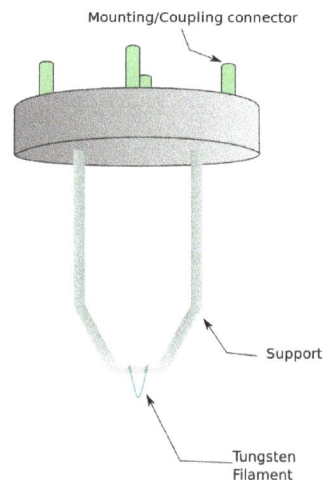

Hairpin style tungsten filament

Manipulation of the electron beam is performed using two physical effects. The interaction of electrons with a magnetic field will cause electrons to move according to the left hand rule, thus allowing for electromagnets to manipulate the electron beam. The use of magnetic fields allows for the formation of a magnetic lens of variable focusing power, the lens shape originating due to the distribution of magnetic flux. Additionally, electrostatic fields can cause the electrons to be deflected through a constant angle. Coupling of two deflections in opposing directions with a small intermediate gap allows for the formation of a shift in the beam path, this being used in TEM for beam shifting, subsequently this is extremely important to STEM. From these two effects, as well as the use of an electron imaging system, sufficient control over the beam path is possible for TEM operation. The optical configuration of a TEM can be rapidly changed, unlike that for an optical microscope, as lenses in the beam path can be enabled, have their strength changed, or be disabled entirely simply via rapid electrical switching, the speed of which is limited by effects such as the magnetic hysteresis of the lenses.

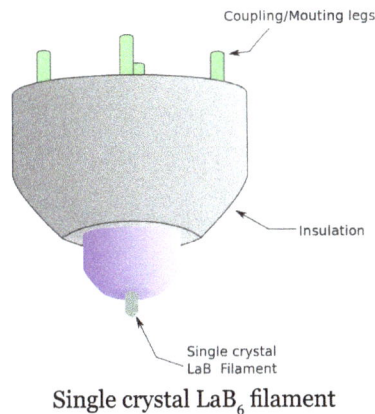

Single crystal LaB$_6$ filament

Optics

The lenses of a TEM allow for beam convergence, with the angle of convergence as a variable parameter, giving the TEM the ability to change magnification simply by modifying the amount of current that flows through the coil, quadrupole or hexapole lenses. The quadrupole lens is an arrangement of electromagnetic coils at the vertices of the square, enabling the generation of a lensing magnetic fields, the hexapole configuration simply enhances the lens symmetry by using six, rather than four coils.

Typically a TEM consists of three stages of lensing. The stages are the condenser lenses, the objective lenses, and the projector lenses. The condenser lenses are responsible for primary beam formation, while the objective lenses focus the beam that comes through the sample itself (in STEM scanning mode, there are also objective lenses above the sample to make the incident electron beam convergent). The projector lenses are used to expand the beam onto the phosphor screen or other imaging device, such as film. The magnification of the TEM is due to the ratio of the distances between the specimen and the objective lens' image plane. Additional stigmators allow for the correction of asymmetrical beam distortions, known as astigmatism. It is noted that TEM optical configurations differ significantly with implementation, with manufacturers using custom lens configurations, such as in spherical aberration corrected instruments, or TEMs utilizing energy filtering to correct electron chromatic aberration.

Display

Imaging systems in a TEM consist of a phosphor screen, which may be made of fine (10–100 µm) particulate zinc sulfide, for direct observation by the operator, and, optionally, an image recording system such as film based or doped YAG screen coupled CCDs. Typically these devices can be removed or inserted into the beam path by the operator as required.

Components

The electron source of the TEM is at the top, where the lensing system (4,7 and 8) focuses the beam on the specimen and then projects it onto the viewing screen (10). The beam control is on the right (13 and 14)

A TEM is composed of several components, which include a vacuum system in which the electrons travel, an electron emission source for generation of the electron stream, a series of electromagnetic lenses, as well as electrostatic plates. The latter two allow the operator to guide and manipulate the beam as required. Also required is a device to allow the insertion into, motion within, and removal of specimens from the beam path. Imaging devices are subsequently used to create an image from the electrons that exit the system.

Vacuum System

To increase the mean free path of the electron gas interaction, a standard TEM is evacuated to low pressures, typically on the order of 10^{-4} Pa. The need for this is twofold: first the allowance for the voltage difference between the cathode and the ground without generating an arc, and secondly to reduce the collision frequency of electrons with gas atoms to negligible levels—this effect is characterized by the mean free path. TEM components such as specimen holders and film cartridges must be routinely inserted or replaced requiring a system with the ability to re-evacuate on a regular basis. As such, TEMs are equipped with multiple pumping systems and airlocks and are not permanently vacuum sealed.

The vacuum system for evacuating a TEM to an operating pressure level consists of several stages. Initially, a low or roughing vacuum is achieved with either a rotary vane pump or diaphragm pumps setting a sufficiently low pressure to allow the operation of a turbo-molecular or diffusion pump establishing high vacuum level necessary for operations. To allow for the low vacuum pump to not require continuous operation, while continually operating the turbo-molecular pumps, the vacuum side of a low-pressure pump may be connected to chambers which accommodate the exhaust gases from the turbo-molecular pump. Sections of the TEM may be isolated by the use of pressure-limiting apertures to allow for different vacuum levels in specific areas such as a higher vacuum of 10^{-4} to 10^{-7} Pa or higher in the electron gun in high-resolution or field-emission TEMs.

High-voltage TEMs require ultra-high vacuums on the range of 10^{-7} to 10^{-9} Pa to prevent the generation of an electrical arc, particularly at the TEM cathode. As such for higher voltage TEMs a third vacuum system may operate, with the gun isolated from the main chamber either by gate valves or a differential pumping aperture – a small hole that prevents the diffusion of gas molecules into the higher vacuum gun area faster than they can be pumped out. For these very low pressures, either an ion pump or a getter material is used.

Poor vacuum in a TEM can cause several problems ranging from the deposition of gas inside the TEM onto the specimen while viewed in a process known as electron beam induced deposition to more severe cathode damages caused by electrical discharge. The use of a cold trap to adsorb sublimated gases in the vicinity of the specimen largely eliminates vacuum problems that are caused by specimen sublimation.

Specimen Stage

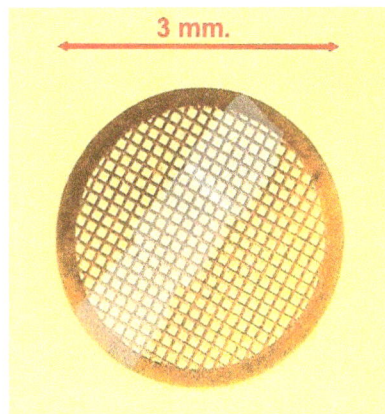

TEM sample support mesh "grid", with ultramicrotomy sections

TEM specimen stage designs include airlocks to allow for insertion of the specimen holder into the vacuum with minimal loss of vacuum in other areas of the microscope. The specimen holders hold a standard size of sample grid or self-supporting specimen. Standard TEM grid sizes are 3.05 mm diameter, with a thickness and mesh size ranging from a few to 100 μm. The sample is placed onto the meshed area having a diameter of approximately 2.5 mm. Usual grid materials are copper, molybdenum, gold or platinum. This grid is placed into the sample holder, which is paired with the specimen stage. A wide variety of designs of stages and holders exist, depending upon the type of experiment being performed. In addition to 3.05 mm grids, 2.3 mm grids are sometimes, if rarely, used. These grids were particularly used in the mineral sciences where a large degree of tilt can be required and where specimen material may be extremely rare. Electron transparent specimens have a thickness usually less than 100 nm, but this value depends on the accelerating voltage.

Once inserted into a TEM, the sample has to be manipulated to locate the region of interest to the beam, such as in single grain diffraction, in a specific orientation. To accommodate this, the TEM stage allows movement of the sample in the XY plane, Z height adjustment, and commonly a single tilt direction parallel to the axis of side entry bolders. Sample rotation may be available on specialized diffraction holders and stages. Some modern TEMs provide the ability for two orthogonal tilt angles of movement with specialized holder designs called double-tilt sample holders. Some stage designs, such as top-entry or vertical insertion stages once common for high resolution TEM studies, may simply only have X-Y translation available. The design criteria of TEM stages are complex, owing to the simultaneous requirements of mechanical and electron-optical constraints and specialized models are available for different methods.

A TEM stage is required to have the ability to hold a specimen and be manipulated to bring the region of interest into the path of the electron beam. As the TEM can operate over a wide range of magnifications, the stage must simultaneously be highly resistant to mechanical drift, with drift requirements as low as a few nm/minute while being able to move several μm/minute, with repositioning accuracy on the order of nanometers. Earlier designs of TEM accomplished this with a complex set of mechanical downgearing devices, allowing the operator to finely control the motion of the stage by several rotating rods. Modern devices may use electrical stage designs, using screw gearing in concert with stepper motors, providing the operator with a computer-based stage input, such as a joystick or trackball.

Two main designs for stages in a TEM exist, the side-entry and top entry version. Each design must accommodate the matching holder to allow for specimen insertion without either damaging delicate TEM optics or allowing gas into TEM systems under vacuum.

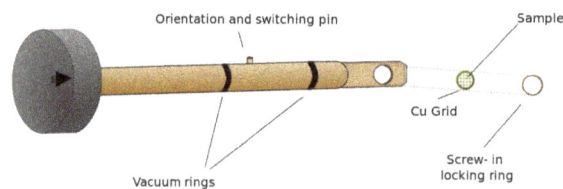

A diagram of a single axis tilt sample holder for insertion into a TEM goniometer.
Titling of the holder is achieved by rotation of the entire goniometer

The most common is the side entry holder, where the specimen is placed near the tip of a long metal (brass or stainless steel) rod, with the specimen placed flat in a small bore. Along the rod are several polymer vacuum rings to allow for the formation of a vacuum seal of sufficient quality, when inserted into the stage. The stage is thus designed to accommodate the rod, placing the sample either in between or near the objective lens, dependent upon the objective design. When inserted into the stage, the side entry holder has its tip contained within the TEM vacuum, and the base is presented to atmosphere, the airlock formed by the vacuum rings.

Insertion procedures for side-entry TEM holders typically involve the rotation of the sample to trigger micro switches that initiate evacuation of the airlock before the sample is inserted into the TEM column.

The second design is the top-entry holder consists of a cartridge that is several cm long with a bore drilled down the cartridge axis. The specimen is loaded into the bore, possibly utilizing a small screw ring to hold the sample in place. This cartridge is inserted into an airlock with the bore perpendicular to the TEM optic axis. When sealed, the airlock is manipulated to push the cartridge such that the cartridge falls into place, where the bore hole becomes aligned with the beam axis, such that the beam travels down the cartridge bore and into the specimen. Such designs are typically unable to be tilted without blocking the beam path or interfering with the objective lens.

Electron Gun

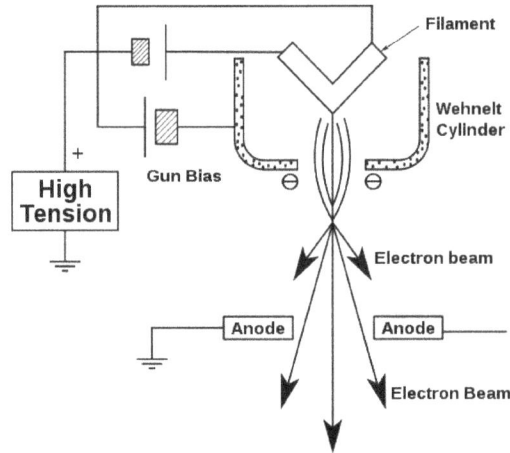

Cross sectional diagram of an electron gun assembly, illustrating electron extraction

The electron gun is formed from several components: the filament, a biasing circuit, a Wehnelt cap, and an extraction anode. By connecting the filament to the negative component power supply, electrons can be "pumped" from the electron gun to the anode plate, and TEM column, thus completing the circuit. The gun is designed to create a beam of electrons exiting from the assembly at some given angle, known as the gun divergence semi-angle, α. By constructing the Wehnelt cylinder such that it has a higher negative charge than the filament itself, electrons that exit the filament in a diverging manner are, under proper operation, forced into a converging pattern the minimum size of which is the gun crossover diameter.

The thermionic emission current density, J, can be related to the work function of the emitting material via Richardson's law

$$J = AT^2 \exp\left(-\frac{\Phi}{kT}\right),$$

where A is the Richardson's constant, Φ is the work function and T is the temperature of the material.

This equation shows that in order to achieve sufficient current density it is necessary to heat the emitter, taking care not to cause damage by application of excessive heat, for this reason materials with either a high melting point, such as tungsten, or those with a low work function (LaB_6) are required for the gun filament. Furthermore, both lanthanum hexaboride and tungsten thermionic sources must be heated in order to achieve thermionic emission, this can be achieved by the use of a small resistive strip. To prevent thermal shock, there is often a delay enforced in the application of current to the tip, to prevent thermal gradients from damaging the filament, the delay is usually a few seconds for LaB_6, and significantly lower for tungsten.

Electron Lens

Electron lenses are designed to act in a manner emulating that of an optical lens, by focusing parallel electrons at some constant focal distance. Electron lenses may operate electrostatically or magnetically. The majority of electron lenses for TEM use electromagnetic coils to generate a convex lens. The field produced for the lens must be radially symmetrical, as deviation from the radial symmetry of the magnetic lens causes aberrations such as astigmatism, and worsens spherical and chromatic aberration. Electron lenses are manufactured from iron, iron-cobalt or nickel cobalt alloys, such as permalloy. These are selected for their magnetic properties, such as magnetic saturation, hysteresis and permeability.

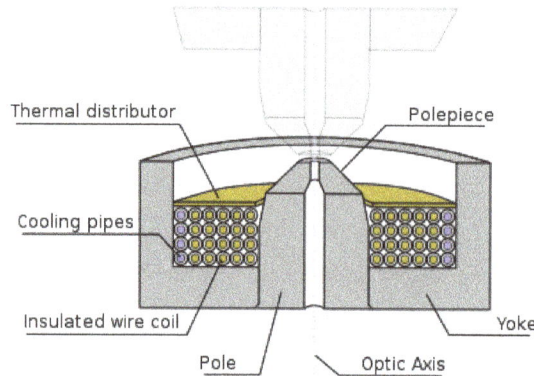

Diagram of a TEM split pole piece design lens

The components include the yoke, the magnetic coil, the poles, the polepiece, and the external control circuitry. The pole piece must be manufactured in a very symmetrical manner, as this provides the boundary conditions for the magnetic field that forms the lens. Imperfections in the manufacture of the pole piece can induce severe distortions in the magnetic field symmetry, which induce distortions that will ultimately limit the lenses' ability to reproduce the object plane. The exact dimensions of the gap, pole piece internal diameter and taper, as well as the overall design of the lens is often performed by finite element analysis of the magnetic field, whilst considering the thermal and electrical constraints of the design.

The coils which produce the magnetic field are located within the lens yoke. The coils can contain a variable current, but typically utilize high voltages, and therefore require significant insulation in order to prevent short-circuiting the lens components. Thermal distributors are placed to ensure the extraction of the heat generated by the energy lost to resistance of the coil windings. The windings may be water-cooled, using a chilled water supply in order to facilitate the removal of the high thermal duty.

Apertures

Apertures are annular metallic plates, through which electrons that are further than a fixed distance from the optic axis may be excluded. These consist of a small metallic disc that is sufficiently thick to prevent electrons from passing through the disc, whilst permitting axial electrons. This permission of central electrons in a TEM causes two effects simultaneously: firstly, apertures decrease the beam intensity as electrons are filtered from the beam, which may be desired in the case of beam sensitive samples. Secondly, this filtering removes electrons that are scattered to high angles, which may be due to unwanted processes such as spherical or chromatic aberration, or due to diffraction from interaction within the sample.

Apertures are either a fixed aperture within the column, such as at the condenser lens, or are a movable aperture, which can be inserted or withdrawn from the beam path, or moved in the plane perpendicular to the beam path. Aperture assemblies are mechanical devices which allow for the selection of different aperture sizes, which may be used by the operator to trade off intensity and the filtering effect of the aperture. Aperture assemblies are often equipped with micrometers to move the aperture, required during optical calibration.

Imaging Methods

Imaging methods in TEM utilize the information contained in the electron waves exiting from the sample to form an image. The projector lenses allow for the correct positioning of this electron wave distribution onto the viewing system. The observed intensity, I, of the image, assuming sufficiently high quality of imaging device, can be approximated as proportional to the time-averaged amplitude of the electron wavefunctions, where the wave that forms the exit beam is denoted by Ψ.

$$I(x) = \frac{k}{t_1 - t_0} \int_{t_0}^{t_1} \Psi\Psi^* dt$$

Different imaging methods therefore attempt to modify the electron waves exiting the sample in a way that provides information about the sample, or the beam itself. From the previous equation, it can be deduced that the observed image depends not only on the amplitude of beam, but also on the phase of the electrons, although phase effects may often be ignored at lower magnifications. Higher resolution imaging requires thinner samples and higher energies of incident electrons, which means that the sample can no longer be considered to be absorbing electrons (i.e., via a Beer's law effect). Instead, the sample can be modeled as an object that does not change the amplitude of the incoming electron wave function, but instead modifies the phase of the incoming wave; in this model, the sample is known as a pure phase object. For sufficiently thin specimens, phase effects dominate the image, complicating analysis of the observed intensities. To improve the contrast in the image, the TEM may be operated at a slight defocus to enhance contrast, owing to convolution by the contrast transfer function of the TEM, which would normally decrease contrast if the sample was not a weak phase object.

Contrast Formation

Contrast formation in the TEM depends upon the mode of operation. These different modes may be selected to discern different types of information about the specimen.

Bright Field

The most common mode of operation for a TEM is the bright field imaging mode. In this mode the contrast formation comes from the sample having varying thickness or density. Thicker areas of the sample and denser areas or regions with a higher atomic number will block more electrons and appear dark in an image, while thinner, lower density, lower atomic number regions and areas with no sample in the beam path will appear bright. The term "bright field" refers to the bright background field where there is no sample and most of the beam electrons reach the image. The image is assumed to be a simple two dimensional projection of the sample's thickness and density down the optic axis, and to a first approximation may be modelled via Beer's law, more complex analyses require the modelling of the sample image to include phase information.

Diffraction Contrast

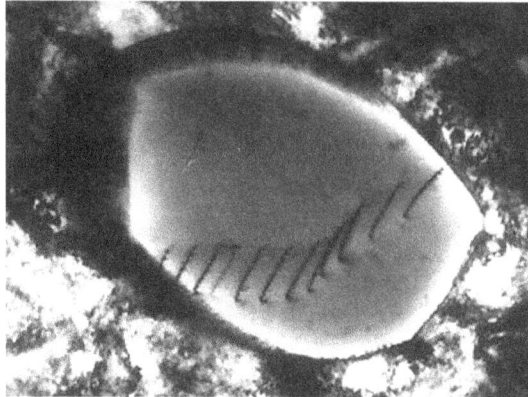

Transmission electron micrograph of dislocations in steel, which are faults
in the structure of the crystal lattice at the atomic scale

Samples can exhibit diffraction contrast, whereby the electron beam undergoes Bragg scattering, which in the case of a crystalline sample, disperses electrons into discrete locations in the back focal plane. By the placement of apertures in the back focal plane, i.e. the objective aperture, the desired Bragg reflections can be selected (or excluded), thus only parts of the sample that are causing the electrons to scatter to the selected reflections will end up projected onto the imaging apparatus.

If the reflections that are selected do not include the unscattered beam (which will appear up at the focal point of the lens), then the image will appear dark wherever no sample scattering to the selected peak is present, as such a region without a specimen will appear dark. This is known as a dark-field image.

Modern TEMs are often equipped with specimen holders that allow the user to tilt the specimen to a range of angles in order to obtain specific diffraction conditions, and apertures placed above the specimen allow the user to select electrons that would otherwise be diffracted in a particular direction from entering the specimen.

Applications for this method include the identification of lattice defects in crystals. By carefully selecting the orientation of the sample, it is possible not just to determine the position of defects but also to determine the type of defect present. If the sample is oriented so that one particular plane is only slightly tilted away from the strongest diffracting angle (known as the Bragg Angle), any distortion of the crystal plane that locally tilts the plane to the Bragg angle will produce particularly strong contrast variations. However, defects that produce only displacement of atoms that do not tilt the crystal to the Bragg angle (i. e. displacements parallel to the crystal plane) will not produce strong contrast.

Electron Energy Loss

Utilizing the advanced technique of EELS, for TEMs appropriately equipped, electrons can be rejected based upon their voltage (which, due to constant charge is their energy), using magnetic sector based devices known as EELS spectrometers. These devices allow for the selection of particular energy values, which can be associated with the way the electron has interacted with the sample.

For example, different elements in a sample result in different electron energies in the beam after the sample. This normally results in chromatic aberration – however this effect can, for example, be used to generate an image which provides information on elemental composition, based upon the atomic transition during electron-electron interaction.

EELS spectrometers can often be operated in both spectroscopic and imaging modes, allowing for isolation or rejection of elastically scattered beams. As for many images inelastic scattering will include information that may not be of interest to the investigator thus reducing observable signals of interest, EELS imaging can be used to enhance contrast in observed images, including both bright field and diffraction, by rejecting unwanted components.

Phase Contrast

Crystal structure can also be investigated by high-resolution transmission electron microscopy (HRTEM), also known as phase contrast. When utilizing a Field emission source and a specimen of uniform thickness, the images are formed due to differences in phase of electron waves, which is caused by specimen interaction. Image formation is given by the complex modulus of the incoming electron beams. As such, the image is not only dependent on the number of electrons hitting the screen, making direct interpretation of phase contrast images more complex. However this effect can be used to an advantage, as it can be manipulated to provide more information about the sample, such as in complex phase retrieval techniques.

Diffraction

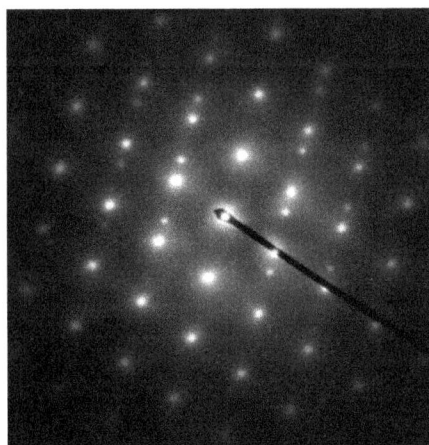

Crystalline diffraction pattern from a twinned grain of FCC Austenitic steel

As previously stated, by adjusting the magnetic lenses such that the back focal plane of the lens rather than the imaging plane is placed on the imaging apparatus a diffraction pattern can be generated. For thin crystalline samples, this produces an image that consists of a pattern of dots in the case of a single crystal, or a series of rings in the case of a polycrystalline or amorphous solid material. For the single crystal case the diffraction pattern is dependent upon the orientation of the specimen and the structure of the sample illuminated by the electron beam. This image provides the investigator with information about the space group symmetries in the crystal and the crystal's orientation to the beam path. This is typically done without utilizing any information but the position at which the diffraction spots appear and the observed image symmetries.

Diffraction patterns can have a large dynamic range, and for crystalline samples, may have intensities greater than those recordable by CCD. As such, TEMs may still be equipped with film cartridges for the purpose of obtaining these images, as the film is a single use detector.

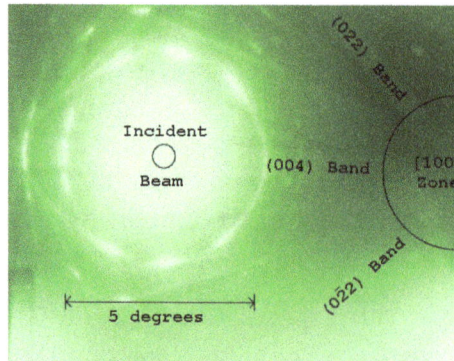

Convergent-beam Kikuchi lines from silicon, near the zone axis

Analysis of diffraction patterns beyond point-position can be complex, as the image is sensitive to a number of factors such as specimen thickness and orientation, objective lens defocus, spherical and chromatic aberration. Although quantitative interpretation of the contrast shown in lattice images is possible, it is inherently complicated and can require extensive computer simulation and analysis, such as electron multislice analysis.

More complex behaviour in the diffraction plane is also possible, with phenomena such as Kikuchi lines arising from multiple diffraction within the crystalline lattice. In convergent beam electron diffraction (CBED) where a non-parallel, i.e. converging, electron wavefront is produced by concentrating the electron beam into a fine probe at the sample surface, the interaction of the convergent beam can provide information beyond structural data such as sample thickness.

Three-dimensional Imaging

As TEM specimen holders typically allow for the rotation of a sample by a desired angle, multiple views of the same specimen can be obtained by rotating the angle of the sample along an axis perpendicular to the beam. By taking multiple images of a single TEM sample at differing angles, typically in 1° increments, a set of images known as a "tilt series" can be collected. This methodology was proposed in the 1970s by Walter Hoppe. Under purely absorption contrast conditions, this set of images can be used to construct a three-dimensional representation of the sample.

The reconstruction is accomplished by a two-step process, first images are aligned to account for errors in the positioning of a sample; such errors can occur due to vibration or mechanical drift. Alignment methods use image registration algorithms, such as autocorrelation methods to correct these errors. Secondly, using a reconstruction algorithm, such as filtered back projection, the aligned image slices can be transformed from a set of two-dimensional images, $I_j(x, y)$, to a single three-dimensional image, $I'_j(x, y, z)$. This three-dimensional image is of particular interest when morphological information is required, further study can be undertaken using computer algorithms, such as isosurfaces and data slicing to analyse the data.

As TEM samples cannot typically be viewed at a full 180° rotation, the observed images typically suffer from a "missing wedge" of data, which when using Fourier-based back projection methods decreases the range of resolvable frequencies in the three-dimensional reconstruction. Mechanical refinements, such as multi-axis tilting (two tilt series of the same specimen made at orthogonal directions) and conical tomography (where the specimen is first tilted to a given fixed angle and then imaged at equal angular rotational increments through one complete rotation in the plane of the specimen grid) can be used to limit the impact of the missing data on the observed specimen morphology. Using focused ion beam milling, a new technique has been proposed which uses pillar-shaped specimen and a dedicated on-axis tomography holder to perform 180° rotation of the sample inside the pole piece of the objective lens in TEM. Using such arrangements, quantitative electron tomography without the missing wedge is possible. In addition, numerical techniques exist which can improve the collected data.

All the above-mentioned methods involve recording tilt series of a given specimen field. This inevitably results in the summation of a high dose of reactive electrons through the sample and the accompanying destruction of fine detail during recording. The technique of low-dose (minimal-dose) imaging is therefore regularly applied to mitigate this effect. Low-dose imaging is performed by deflecting illumination and imaging regions simultaneously away from the optical axis to image an adjacent region to the area to be recorded (the high-dose region). This area is maintained centered during tilting and refocused before recording. During recording the deflections are removed so that the area of interest is exposed to the electron beam only for the duration required for imaging. An improvement of this technique (for objects resting on a sloping substrate film) is to have two symmetrical off-axis regions for focusing followed by setting focus to the average of the two high-dose focus values before recording the low-dose area of interest.

Non-tomographic variants on this method, referred to as single particle analysis, use images of multiple (hopefully) identical objects at different orientations to produce the image data required for three-dimensional reconstruction. If the objects do not have significant preferred orientations, this method does not suffer from the missing data wedge (or cone) which accompany tomographic methods nor does it incur excessive radiation dosage, however it assumes that the different objects imaged can be treated as if the 3D data generated from them arose from a single stable object.

Sample Preparation

A sample of cells (black) stained with osmium tetroxide and uranyl acetate
embedded in epoxy resin (amber) ready for sectioning.

Sample preparation in TEM can be a complex procedure. TEM specimens should be less than 100 nanometers thick for a conventional TEM. Unlike neutron or X-Ray radiation the electrons in the beam interact readily with the sample, an effect that increases roughly with atomic number squared (z^2). High quality samples will have a thickness that is comparable to the mean free path of the electrons that travel through the samples, which may be only a few tens of nanometers. Preparation of TEM specimens is specific to the material under analysis and the type of information to be obtained from the specimen.

Materials that have dimensions small enough to be electron transparent, such as powdered substances, small organisms, viruses, or nanotubes, can be quickly prepared by the deposition of a dilute sample containing the specimen onto films on support grids. Biological specimens may be embedded in resin to withstand the high vacuum in the sample chamber and to enable cutting tissue into electron transparent thin sections. The biological sample can be stained using either a negative staining material such as uranyl acetate for bacteria and viruses, or, in the case of embedded sections, the specimen may be stained with heavy metals, including osmium tetroxide. Alternately samples may be held at liquid nitrogen temperatures after embedding in vitreous ice. In material science and metallurgy the specimens can usually withstand the high vacuum, but still must be prepared as a thin foil, or etched so some portion of the specimen is thin enough for the beam to penetrate. Constraints on the thickness of the material may be limited by the scattering cross-section of the atoms from which the material is comprised.

Tissue Sectioning

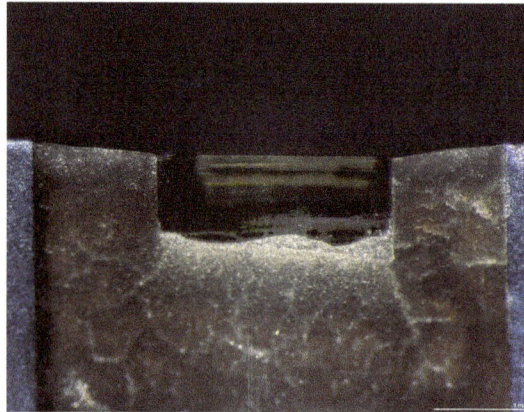

A diamond knife blade used for cutting ultrathin sections (typically 70 to 350 nm) for transmission electron microscopy.

Biological tissue is often embedded in a resin block then thinned to less than 100nm on an ultramicrotome. The resin block is fractured as it passes over a glass or diamond knife edge. This method is used to obtain thin, minimally deformed samples that allow for the observation of tissue ultrastructure. Inorganic samples, such as aluminium, may also be embedded in resins and ultrathin sectioned in this way, using either coated glass, sapphire or larger angle diamond knives. To prevent charge build-up at the sample surface when viewing in the TEM, tissue samples need to be coated with a thin layer of conducting material, such as carbon.

Sample Staining

TEM samples of biological tissues need high atomic number stains to enhance contrast. The stain absorbs the beam electrons or scatters part of the electron beam which otherwise is projected onto the imaging system. Compounds of heavy metals such as osmium, lead, uranium or gold (in immunogold labelling) may be used prior to TEM observation to selectively deposit electron dense atoms in or on the sample in desired cellular or protein regionn. This process requires an understanding of how heavy metals bind to specific biological tissues and cellular structures.

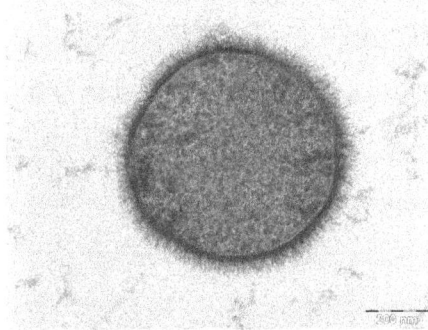

A section of a cell of *Bacillus subtilis*, taken with a Tecnai T-12 TEM. The scale bar is 200 nm.

Mechanical Milling

Mechanical polishing is also used to prepare samples for imaging on the TEM. Polishing needs to be done to a high quality, to ensure constant sample thickness across the region of interest. A diamond, or cubic boron nitride polishing compound may be used in the final stages of polishing to remove any scratches that may cause contrast fluctuations due to varying sample thickness. Even after careful mechanical milling, additional fine methods such as ion etching may be required to perform final stage thinning.

Chemical Etching

Certain samples may be prepared by chemical etching, particularly metallic specimens. These samples are thinned using a chemical etchant, such as an acid, to prepare the sample for TEM observation. Devices to control the thinning process may allow the operator to control either the voltage or current passing through the specimen, and may include systems to detect when the sample has been thinned to a sufficient level of optical transparency.

Ion Etching

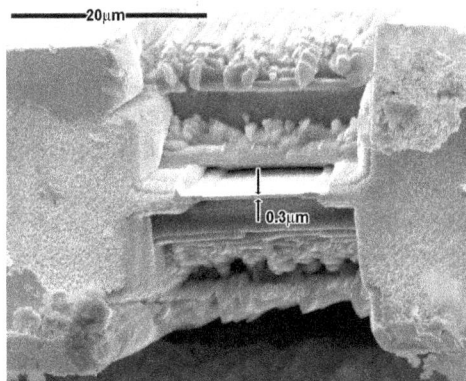

SEM image of a thin TEM sample milled by FIB. The thin membrane shown here is suitable for TEM examination; however, at ~300-nm thickness, it would not be suitable for high-resolution TEM without further milling.

Ion etching is a sputtering process that can remove very fine quantities of material. This is used to perform a finishing polish of specimens polished by other means. Ion etching uses an inert gas passed through an electric field to generate a plasma stream that is directed to the sample surface. Acceleration energies for gases such as argon are typically a few kilovolts. The sample may be rotated to promote even polishing of the sample surface. The sputtering rate of such methods is on the order of tens of micrometers per hour, limiting the method to only extremely fine polishing.

Ion etching by argon gas has been recently shown to be able to file down MTJ stack structures to a specific layer which has then been atomically resolved. The TEM images taken in plan view rather than cross-section reveal that the MgO layer within MTJs contains a large number of grain boundaries that may be diminishing the properties of devices.

Ion Milling

More recently focused ion beam methods have been used to prepare samples. FIB is a relatively new technique to prepare thin samples for TEM examination from larger specimens. Because FIB can be used to micro-machine samples very precisely, it is possible to mill very thin membranes from a specific area of interest in a sample, such as a semiconductor or metal. Unlike inert gas ion sputtering, FIB makes use of significantly more energetic gallium ions and may alter the composition or structure of the material through gallium implantation.

Replication

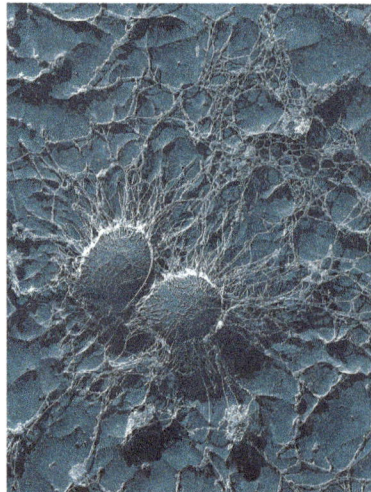

Staphylococcus aureus platinum replica image shot on a TEM at 50,000x magnification

Samples may also be replicated using cellulose acetate film, the film subsequently coated with a heavy metal such as platinum, the original film dissolved away, and the replica imaged on the TEM. Variations of the replica technique are used for both materials and biological samples. In materials science a common use is for examining the fresh fracture surface of metal alloys.

Modifications

The capabilities of the TEM can be further extended by additional stages and detectors, sometimes incorporated on the same microscope.

Scanning TEM

A TEM can be modified into a scanning transmission electron microscope (STEM) by the addition of a system that rasters the beam across the sample to form the image, combined with suitable detectors. Scanning coils are used to deflect the beam, such as by an electrostatic shift of the beam, where the beam is then collected using a current detector such as a Faraday cup, which acts as a direct electron counter. By correlating the electron count to the position of the scanning beam (known as the "probe"), the transmitted component of the beam may be measured. The non-transmitted components may be obtained either by beam tilting or by the use of annular dark field detectors.

Low-voltage Electron Microscope

A low-voltage electron microscope (LVEM) is operated at relatively low electron accelerating voltage between 5–25 kV. Some of these can be a combination of SEM, TEM and STEM in a single compact instrument. Low voltage increases image contrast which is especially important for biological specimens. This increase in contrast significantly reduces, or even eliminates the need to stain. Resolutions of a few nm are possible in TEM, SEM and STEM modes. The low energy of the electron beam means that permanent magnets can be used as lenses and thus a miniature column that does not require cooling can be used.

Cryo-TEM

Cryogenic transmission electron microscopy (Cryo-TEM) uses a TEM with a specimen holder capable of maintaining the specimen at liquid nitrogen or liquid helium temperatures. This allows imaging specimens prepared in vitreous ice, the preferred preparation technique for imaging individual molecules or macromolecular assemblies, imaging of vitrified solid-electrolye interfaces, and imaging of materials that are volatile in high vacuum at room temperature, such as sulfur.

Environmental/In-situ TEM

In-situ experiments may also be conducted in TEM using differentially pumped sample chambers, or specialized holders. Types of in-situ experiments include studying chemical reactions in liquid cells, and material deformation testing.

Aberration Corrected TEM

Modern research TEMs may include aberration correctors, to reduce the amount of distortion in the image. Incident beam monochromators may also be used which reduce the energy spread of the incident electron beam to less than 0.15 eV. Major aberration corrected TEM manufacturers include JEOL, Hitachi High-technologies, FEI Company, and NION.

Limitations

There are a number of drawbacks to the TEM technique. Many materials require extensive sample preparation to produce a sample thin enough to be electron transparent, which makes TEM analysis a relatively time consuming process with a low throughput of samples. The structure of the sample may also be changed during the preparation process. Also the field of view is relatively small, raising the possibility that the region analyzed may not be characteristic of the whole sample. There is potential that the sample may be damaged by the electron beam, particularly in the case of biological materials.

Resolution Limits

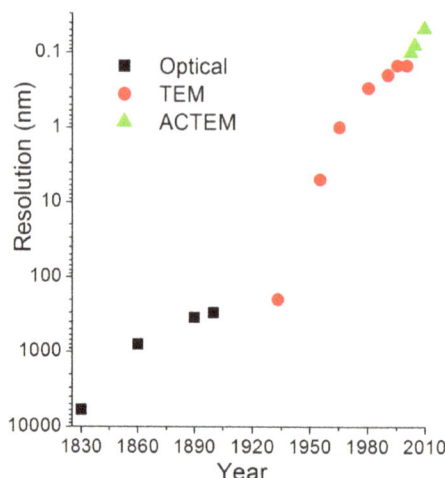

Evolution of spatial resolution achieved with optical, transmission (TEM) and aberration-corrected electron microscopes (ACTEM).

The limit of resolution obtainable in a TEM may be described in several ways, and is typically referred to as the information limit of the microscope. One commonly used value is a cut-off value of the contrast transfer function, a function that is usually quoted in the frequency domain to define the reproduction of spatial frequencies of objects in the object plane by the microscope optics. A cut-off frequency, q_{max}, for the transfer function may be approximated with the following equation, where C_s is the spherical aberration coefficient and λ is the electron wavelength:

$$q_{max} = \frac{1}{0.67(C_s\lambda^3)^{1/4}}.$$

For a 200 kV microscope, with partly corrected spherical aberrations ("to the third order") and a C_s value of 1 μm, a theoretical cut-off value might be $1/q_{max}$ = 42 pm. The same microscope without a corrector would have C_s = 0.5 mm and thus a 200-pm cut-off. The spherical aberrations are suppressed to the third or fifth order in the "aberration-corrected" microscopes. Their resolution is however limited by electron source geometry and brightness and chromatic aberrations in the objective lens system.

The frequency domain representation of the contrast transfer function may often have an oscillatory nature, which can be tuned by adjusting the focal value of the objective lens. This oscillatory nature implies that some spatial frequencies are faithfully imaged by the microscope, whilst others are suppressed. By combining multiple images with different spatial frequencies, the use of techniques such as focal series reconstruction can be used to improve the resolution of the TEM in a limited manner. The contrast transfer function can, to some extent, be experimentally approximated through techniques such as Fourier transforming images of amorphous material, such as amorphous carbon.

More recently, advances in aberration corrector design have been able to reduce spherical aberrations and to achieve resolution below 0.5 Ångströms (50 pm) at magnifications above 50 million times. Improved resolution allows for the imaging of lighter atoms that scatter electrons less efficiently, such as lithium atoms in lithium battery materials. The ability to determine the position of atoms within materials has made the HRTEM an indispensable tool for nanotechnology research and development in many fields, including heterogeneous catalysis and the development of semiconductor devices for electronics and photonics.

Electrons moving at high speed also behave as an electromagnetic wave. Its wave length is given by $\lambda = \sqrt{\dfrac{150}{V}}$ Angstrom, where V is the voltage in kV applied to accelerate the beam of electrons.

Using this if the wavelength of electron beam is calculated for a 100kV electron microscope it comes out to be around 0.04 Angstrom. This is of the order of atomic spacing in crystals. Therefore resolving power of TEM should be extremely high. In principle it should be possible have atomic level resolution. However because of problems associated with the design of electromagnetic lens free from aberrations it has so far been a challenging task.

The requirement of good quality transparent specimen is the key to exploit full potential of TEM. For examination of metallic samples the most commonly used techniques are extraction carbon replica, electrolytic polishing and ion milling. Carbon replica is prepared by vacuum deposition of carbon film on previously polished and lightly etched specimen. This is later stripped by etching. The structural details get transferred to the thin carbon film. Sometimes precipitates get attached to the film during stripping. This allows selected area diffraction pattern to be recorded. It helps in identification of precipitates. Electrolytic polishing or ion milling allows complete examination of the metallic samples. All these need excellent experimental skill and a lot of practice. In this case the initial sample should be thin enough.

Scanning Electron Microscope

A scanning electron microscope (SEM) is a type of electron microscope that produces images of a sample by scanning the surface with a focused beam of electrons. The electrons interact with atoms in the sample, producing various signals that contain information about the sample's surface topography and composition. The electron beam is scanned in a raster scan pattern, and the beam's position is combined with the detected signal to produce an image. SEM can achieve resolution better than 1 nanometer. Specimens can be observed in high vacuum in conventional SEM, or in low vacuum or wet conditions in variable pressure or environmental SEM, and at a wide range of cryogenic or elevated temperatures with specialized instruments.

The most common SEM mode is detection of secondary electrons emitted by atoms excited by the electron beam. The number of secondary electrons that can be detected depends, among other things, on specimen topography. By scanning the sample and collecting the secondary electrons that are emitted using a special detector, an image displaying the topography of the surface is created.

History

An account of the early history of SEM has been presented by McMullan. Although Max Knoll produced a photo with a 50 mm object-field-width showing channeling contrast by the use of an electron beam scanner, it was Manfred von Ardenne who in 1937 invented a true microscope with high magnification by scanning a very small raster with a demagnified and finely focused electron beam. Ardenne applied the scanning principle not only to achieve magnification but also to purposefully eliminate the chromatic aberration otherwise inherent in the electron microscope. He further discussed the various detection modes, possibilities and theory of SEM, together with the construction of the first high magnification SEM. Further work was reported by Zworykin's group, followed by the Cambridge groups in the 1950s and early 1960s headed by Charles Oatley, all of which finally led to the marketing of the first commercial instrument by Cambridge Scientific Instrument Company as the "Stereoscan" in 1965, which was delivered to DuPont.

Principles and Capacities

The signals used by a scanning electron microscope to produce an image result from interactions of the electron beam with atoms at various depths within the sample. Various types of signals are produced including secondary electrons (SE), reflected or back-scattered electrons (BSE), characteristic X-rays and light (cathodoluminescence) (CL), absorbed current (specimen current) and transmitted electrons. Secondary electron detectors are standard equipment in all SEMs, but it is rare that a single machine would have detectors for all other possible signals.

In secondary electron imaging, or SEI, the secondary electrons are emitted from very close to the specimen surface. Consequently, SEM can produce very high-resolution images of a sample surface, revealing details less than 1 nm in size. Back-scattered electrons (BSE) are beam electrons that are reflected from the sample by elastic scattering. They emerge from deeper locations within the specimen and consequently the resolution of BSE images is less than SE images. However, BSE are often used in analytical SEM along with the spectra made from the characteristic X-rays, because the intensity of the BSE signal is strongly related to the atomic number (Z) of the specimen. BSE images can provide information about the distribution of different elements in the sample. For the same reason, BSE imaging can image colloidal gold immuno-labels of 5 or 10 nm diameter, which would otherwise be difficult or impossible to detect in secondary electron images in biological specimens. Characteristic X-rays are emitted when the electron beam removes an inner shell electron from the sample, causing a higher-energy electron to fill the shell and release energy. These characteristic X-rays are used to identify the composition and measure the abundance of elements in the sample.

Due to the very narrow electron beam, SEM micrographs have a large depth of field yielding a characteristic three-dimensional appearance useful for understanding the surface structure of a sample. A wide range of magnifications is possible, from about 10 times (about equivalent to that of a powerful hand-lens) to more than 500,000 times, about 250 times the magnification limit of the best light microscopes.

Sample Preparation

Samples for SEM have to be prepared to withstand the vacuum conditions and high energy beam of electrons, and have to be of a size that will fit on the specimen stage. Samples are generally mounted rigidly to a specimen holder or stub using a conductive adhesive. SEM is used extensively for defect analyis of semiconductor wafers, and manufacturers make instruments that can examine any part of a 300 mm semiconductor wafer. Many instruments have chambers that can tilt an object of that size to 45° and provide continuous 360° rotation.

A spider sputter-coated in gold, having been prepared for viewing with an SEM.

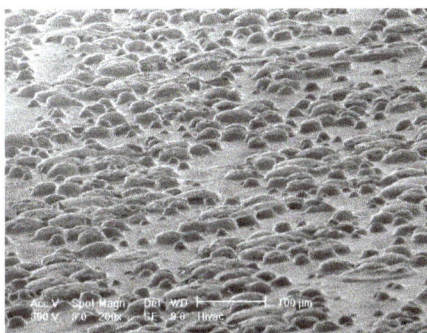

Low-voltage micrograph (300 V) of distribution of adhesive droplets on a Post-it note. No conductive coating was applied: such a coating would alter this fragile specimen.

Nonconductive specimens collect charge when scanned by the electron beam, and especially in secondary electron imaging mode, this causes scanning faults and other image artifacts. For conventional imaging in the SEM, specimens must be electrically conductive, at least at the surface, and electrically grounded to prevent the accumulation of electrostatic charge. Metal objects require little special preparation for SEM except for cleaning and conductively mounting to a specimen stub. Non-conducting materials are usually coated with an ultrathin coating of electrically conducting material, deposited on the sample either by low-vacuum sputter coating or by high-vacuum evaporation. Conductive materials in current use for specimen coating include gold, gold/palladium alloy, platinum, iridium, tungsten, chromium, osmium, and graphite. Coating with heavy metals may increase signal/noise ratio for samples of low atomic number (Z). The improvement arises because secondary electron emission for high-Z materials is enhanced.

An alternative to coating for some biological samples is to increase the bulk conductivity of the material by impregnation with osmium using variants of the OTO staining method (O-osmium tetroxide, T-thiocarbohydrazide, O-osmium).

Nonconducting specimens may be imaged without coating using an environmental SEM (ESEM) or low-voltage mode of SEM operation. In ESEM instruments the specimen is placed in a relatively high-pressure chamber and the electron optical column is differentially pumped to keep vacuum adequately low at the electron gun. The high-pressure region around the sample in the ESEM neutralizes charge and provides an amplification of the secondary electron signal. Low-voltage SEM is typically conducted in an FEG-SEM because field emission guns (FEG) are capable of producing high primary electron brightness and small spot size even at low accelerating potentials. To prevent charging of non-conductive specimens, operating conditions must be adjusted such that the incoming beam current is equal to sum of outcoming secondary and backscattered electrons currents a condition that is more often met at accelerating voltages of 0.3–4 kV.

Synthetic replicas can be made to avoid the use of original samples when they are not suitable or available for SEM examination due to methodological obstacles or legal issues. This technique is achieved in two steps: (1) a mold of the original surface is made using a silicone-based dental elastomer, and (2) a replica of the original surface is obtained by pouring a synthetic resin into the mold.

Embedding in a resin with further polishing to a mirror-like finish can be used for both biological and materials specimens when imaging in backscattered electrons or when doing quantitative X-ray microanalysis.

The main preparation techniques are not required in the environmental SEM outlined below, but some biological specimens can benefit from fixation.

Biological Samples

For SEM, a specimen is normally required to be completely dry, since the specimen chamber is at high vacuum. Hard, dry materials such as wood, bone, feathers, dried insects, or shells (including egg shells) can be examined with little further treatment, but living cells and tissues and whole, soft-bodied organisms require chemical fixation to preserve and stabilize their structure.

Fixation is usually performed by incubation in a solution of a buffered chemical fixative, such as glutaraldehyde, sometimes in combination with formaldehyde and other fixatives, and optionally followed by postfixation with osmium tetroxide. The fixed tissue is then dehydrated. Because air-drying causes collapse and shrinkage, this is commonly achieved by replacement of water in the cells with organic solvents such as ethanol or acetone, and replacement of these solvents in turn with a transitional fluid such as liquid carbon dioxide by critical point drying. The carbon dioxide is finally removed while in a supercritical state, so that no gas–liquid interface is present within the sample during drying.

The dry specimen is usually mounted on a specimen stub using an adhesive such as epoxy resin or electrically conductive double-sided adhesive tape, and sputter-coated with gold or gold/palladium alloy before examination in the microscope. Samples may be sectioned (with a microtome) if information about the organism's internal ultrastructure is to be exposed for imaging.

If the SEM is equipped with a cold stage for cryo microscopy, cryofixation may be used and low-temperature scanning electron microscopy performed on the cryogenically fixed specimens. Cryo-fixed specimens may be cryo-fractured under vacuum in a special apparatus to reveal internal structure, sputter-coated and transferred onto the SEM cryo-stage while still frozen. Low-temperature scanning electron microscopy (LT-SEM) is also applicable to the imaging of temperature-sensitive materials such as ice and fats.

Freeze-fracturing, freeze-etch or freeze-and-break is a preparation method particularly useful for examining lipid membranes and their incorporated proteins in "face on" view. The preparation method reveals the proteins embedded in the lipid bilayer.

Materials

Back-scattered electron imaging, quantitative X-ray analysis, and X-ray mapping of specimens often requires grinding and polishing the surfaces to an ultra smooth surface. Specimens that undergo WDS or EDS analysis are often carbon-coated. In general, metals are not coated prior to imaging in the SEM because they are conductive and provide their own pathway to ground.

Fractography is the study of fractured surfaces that can be done on a light microscope or, commonly, on an SEM. The fractured surface is cut to a suitable size, cleaned of any organic residues, and mounted on a specimen holder for viewing in the SEM.

Integrated circuits may be cut with a focused ion beam (FIB) or other ion beam milling instrument for viewing in the SEM. The SEM in the first case may be incorporated into the FIB.

Metals, geological specimens, and integrated circuits all may also be chemically polished for viewing in the SEM.

Special high-resolution coating techniques are required for high-magnification imaging of inorganic thin films.

Scanning Process and Image Formation

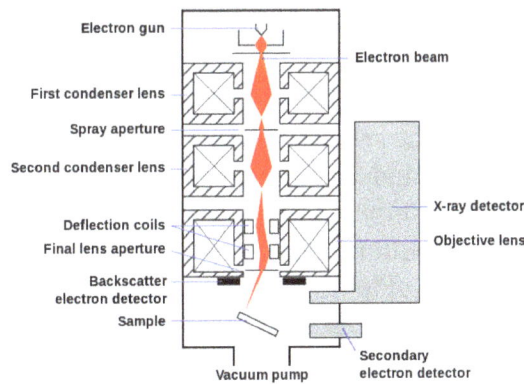

Schematic of an SEM.

In a typical SEM, an electron beam is thermionically emitted from an electron gun fitted with a tungsten filament cathode. Tungsten is normally used in thermionic electron guns because it has the highest melting point and lowest vapor pressure of all metals, thereby allowing it to be electrically heated for electron emission, and because of its low cost. Other types of electron emitters include lanthanum hexaboride (LaB_6) cathodes, which can be used in a standard tungsten filament SEM if the vacuum system is upgraded or field emission guns (FEG), which may be of the cold-cathode type using tungsten single crystal emitters or the thermally assisted Schottky type, that use emitters of zirconium oxide.

The electron beam, which typically has an energy ranging from 0.2 keV to 40 keV, is focused by one or two condenser lenses to a spot about 0.4 nm to 5 nm in diameter. The beam passes through pairs of scanning coils or pairs of deflector plates in the electron column, typically in the final lens, which deflect the beam in the x and y axes so that it scans in a raster fashion over a rectangular area of the sample surface.

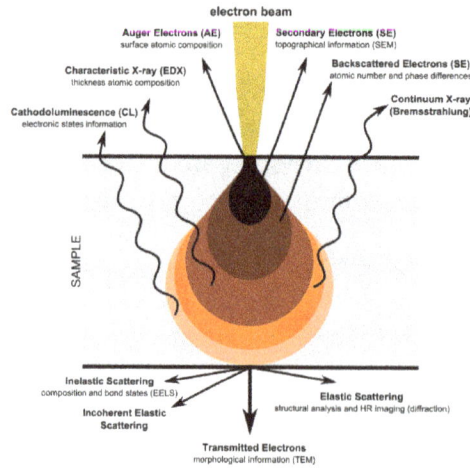

Signals emitted from different parts of the interaction volume

When the primary electron beam interacts with the sample, the electrons lose energy by repeated random scattering and absorption within a teardrop-shaped volume of the specimen known as the interaction volume, which extends from less than 100 nm to approximately 5 µm into the surface. The size of the interaction volume depends on the electron's landing energy, the atomic number of the specimen and the specimen's density. The energy exchange between the electron beam and the sample results in the reflection of high-energy electrons by elastic scattering, emission of secondary electrons by inelastic scattering and the emission of electromagnetic radiation, each of which can be detected by specialized detectors. The beam current absorbed by the specimen can also be detected and used to create images of the distribution of specimen current. Electronic amplifiers of various types are used to amplify the signals, which are displayed as variations in brightness on a computer monitor (or, for vintage models, on a cathode ray tube). Each pixel of computer video memory is synchronized with the position of the beam on the specimen in the microscope, and the resulting image is therefore a distribution map of the intensity of the signal being emitted from the scanned area of the specimen. In older microscopes images may be captured by photography from a high-resolution cathode ray tube, but in modern machines they are digitised and saved as digital images.

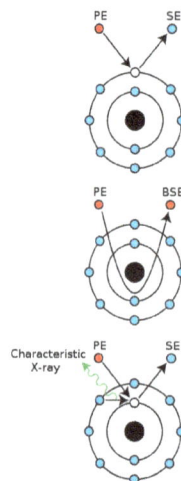

Mechanisms of emission of secondary electrons, backscattered electrons, and characteristic X-rays from atoms of the sample

Magnification

Magnification in a SEM can be controlled over a range of about 6 orders of magnitude from about 10 to 500,000 times. Unlike optical and transmission electron microscopes, image magnification in an SEM is not a function of the power of the objective lens. SEMs may have condenser and objective lenses, but their function is to focus the beam to a spot, and not to image the specimen. Provided the electron gun can generate a beam with sufficiently small diameter, an SEM could in principle work entirely without condenser or objective lenses, although it might not be very versatile or achieve very high resolution. In an SEM, as in scanning probe microscopy, magnification results from the ratio of the dimensions of the raster on the specimen and the raster on the display device. Assuming that the display screen has a fixed size, higher magnification results from reducing the size of the raster on the specimen, and vice versa. Magnification is therefore controlled by the current supplied to the x, y scanning coils, or the voltage supplied to the x, y deflector plates, and not by objective lens power.

Detection of Secondary Electrons

The most common imaging mode collects low-energy (<50 eV) secondary electrons that are ejected from the k-shell of the specimen atoms by inelastic scattering interactions with beam electrons. Due to their low energy, these electrons originate within a few nanometers from the sample surface. The electrons are detected by an Everhart-Thornley detector, which is a type of scintillator-photomultiplier system. The secondary electrons are first collected by attracting them towards an electrically biased grid at about +400 V, and then further accelerated towards a phosphor or scintillator positively biased to about +2,000 V. The accelerated secondary electrons are now sufficiently energetic to cause the scintillator to emit flashes of light (cathodoluminescence), which are conducted to a photomultiplier outside the SEM column via a light pipe and a window in the wall of the specimen chamber. The amplified electrical signal output by the photomultiplier is displayed as a two-dimensional intensity distribution that can be viewed and photographed on an analogue video display, or subjected to analog-to-digital conversion and displayed and saved as a digital image. This process relies on a raster-scanned primary beam. The brightness of the signal depends on the number of secondary electrons reaching the detector. If the beam enters the sample perpendicular to the surface, then the activated region is uniform about the axis of the beam and a certain number of electrons "escape" from within the sample. As the angle of incidence increases, the interaction volume increases and the "escape" distance of one side of the beam decreases, resulting in more secondary electrons being emitted from the sample. Thus steep surfaces and edges tend to be brighter than flat surfaces, which results in images with a well-defined, three-dimensional appearance. Using the signal of secondary electrons image resolution less than 0.5 nm is possible.

Detection of Backscattered Electrons

Backscattered electrons (BSE) consist of high-energy electrons originating in the electron beam, that are reflected or back-scattered out of the specimen interaction volume by elastic scattering interactions with specimen atoms. Since heavy elements (high atomic number) backscatter electrons more strongly than light elements (low atomic number), and thus appear brighter in the image, BSE are used to detect contrast between areas with different chemical compositions. The Everhart-Thornley detector, which is normally positioned to one side of the specimen, is inefficient for the detection of backscattered electrons because few such electrons are emitted in the solid angle subtended by the detector, and because the positively biased detection grid has little ability to attract the higher energy BSE. Dedicated backscattered electron detectors are positioned above the sample in a "doughnut" type arrangement, concentric with the electron beam, maximizing the solid angle of collection. BSE detectors are usually either of scintillator or of semiconductor types. When all parts of the detector are used to collect electrons symmetrically about the beam, atomic number contrast is produced. However, strong topographic contrast is produced by collecting back-scattered electrons from one side above the specimen using an asymmetrical, directional BSE detector; the resulting contrast appears as illumination of the topography from that side. Semiconductor detectors can be made in radial segments that can be switched in or out to control the type of contrast produced and its directionality.

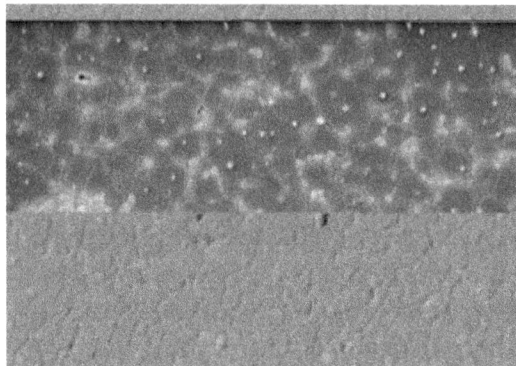

Comparison of SEM techniques:
Top: backscattered electron analysis – composition
Bottom: secondary electron analysis – topography

Backscattered electrons can also be used to form an electron backscatter diffraction (EBSD) image that can be used to determine the crystallographic structure of the specimen.

Beam-injection Analysis of Semiconductors

The nature of the SEM's probe, energetic electrons, makes it uniquely suited to examining the optical and electronic properties of semiconductor materials. The high-energy electrons from the SEM beam will inject charge carriers into the semiconductor. Thus, beam electrons lose energy by promoting electrons from the valence band into the conduction band, leaving behind holes.

In a direct bandgap material, recombination of these electron-hole pairs will result in cathodoluminescence; if the sample contains an internal electric field, such as is present at a p-n junction, the SEM beam injection of carriers will cause electron beam induced current (EBIC) to flow. Cathodoluminescence and EBIC are referred to as "beam-injection" techniques, and are very powerful probes of the optoelectronic behavior of semiconductors, in particular for studying nanoscale features and defects.

Cathodoluminescence

Cathodoluminescence, the emission of light when atoms excited by high-energy electrons return to their ground state, is analogous to UV-induced fluorescence, and some materials such as zinc sulfide and some fluorescent dyes, exhibit both phenomena. Over the last decades, cathodoluminescence was most commonly experienced as the light emission from the inner surface of the cathode ray tube in television sets and computer CRT monitors. In the SEM, CL detectors either collect all light emitted by the specimen or can analyse the wavelengths emitted by the specimen and display an emission spectrum or an image of the distribution of cathodoluminescence emitted by the specimen in real color.

X-ray Microanalysis

Characteristic X-rays that are produced by the interaction of electrons with the sample may also be detected in an SEM equipped for energy-dispersive X-ray spectroscopy or wavelength dispersive X-ray spectroscopy. Analysis of the x-ray signals may be used to map the distribution and estimate the abundance of elements in the sample.

Resolution of the SEM

SEM is not a camera and the detector is not continuously image-forming like a CCD array or film. Unlike in an optical system, the resolution is not limited by the diffraction limit, fineness of lenses or mirrors or detector array resolution. The focusing optics can be large and coarse, and the SE detector is fist-sized and simply detects current. Instead, the spatial resolution of the SEM depends on the size of the electron spot, which in turn depends on both the wavelength of the electrons and the electron-optical system that produces the scanning beam. The resolution is also limited by the size of the interaction volume, the volume of specimen material that interacts with the electron beam. The spot size and the interaction volume are both large compared to the distances between atoms, so the resolution of the SEM is not high enough to image individual atoms, as is possible with transmission electron microscope (TEM). The SEM has compensating advantages, though, including the ability to image a comparatively large area of the specimen; the ability to image bulk materials (not just thin films or foils); and the variety of analytical modes available for measuring the composition and properties of the specimen. Depending on the instrument, the resolution can fall somewhere between less than 1 nm and 20 nm. As of 2009, The world's highest resolution conventional (<30 kV) SEM can reach a point resolution of 0.4 nm using a secondary electron detector.

Environmental SEM

Conventional SEM requires samples to be imaged under vacuum, because a gas atmosphere rapidly spreads and attenuates electron beams. As a consequence, samples that produce a significant amount of vapour, e.g. wet biological samples or oil-bearing rock, must be either dried or cryogenically frozen. Processes involving phase transitions, such as the drying of adhesives or melting of alloys, liquid transport, chemical reactions, and solid-air-gas systems, in general cannot be observed. Some observations of living insects have been possible however.

The first commercial development of the ESEM in the late 1980s allowed samples to be observed in low-pressure gaseous environments (e.g. 1–50 Torr or 0.1–6.7 kPa) and high relative humidity (up to 100%). This was made possible by the development of a secondary-electron detector capable of operating in the presence of water vapour and by the use of pressure-limiting apertures with differential pumping in the path of the electron beam to separate the vacuum region (around the gun and lenses) from the sample chamber.

The first commercial ESEMs were produced by the ElectroScan Corporation in USA in 1988. ElectroScan was taken over by Philips (who later sold their electron-optics division to FEI Company) in 1996.

ESEM is especially useful for non-metallic and biological materials because coating with carbon or gold is unnecessary. Uncoated Plastics and Elastomers can be routinely examined, as can uncoated biological samples. Coating can be difficult to reverse, may conceal small features on the surface of the sample and may reduce the value of the results obtained. X-ray analysis is difficult with a coating of a heavy metal, so carbon coatings are routinely used in conventional SEMs, but ESEM makes it possible to perform X-ray microanalysis on uncoated non-conductive specimens; however er some specific for ESEM artifacts are introduced in X-ray analysis. ESEM may be the preferred for electron microscopy of unique samples from criminal or civil actions, where forensic analysis may need to be repeated by several different experts.

Transmission SEM

The SEM can also be used in transmission mode by simply incorporating an appropriate detector below a thin specimen section . Both bright and dark field imaging has been reported in the generally low accelerating beam voltage range used in SEM, which increases the contrast of unstained biological specimens at high magnifications with a field emission electron gun. This mode of operation has been abbreviated by the acronym tSEM.

Color in SEM

Electron microscopes do not naturally produce color images, as an SEM produces a single value per pixel; this value corresponds to the number of electrons received by the detector during a small period of time of the scanning when the beam is targeted to the (x,y) pixel position.

This single number is usually represented, for each pixel, by a grey level, forming a "black-and-white" image. However, several ways have been used to get color electron microscopy images.

False Color using a Single Detector

- On compositional images of flat surfaces (typically BSE):

The easiest way to get color is to associate to this single number an arbitrary color, using a color look-up table (i.e. each grey level is replaced by a chosen color). This method is known as false color. On a BSE image, false color may be performed to better distinguish the various phases of the sample.

- On textured-surface images:

As an alternative to simply replacing each grey level by a color, a sample observed by an oblique beam allows to create an approximative topography image. Such topography can then be processed by 3D-rendering algorithms for a more natural rendering of the surface texture.

Surface of a kidney stone

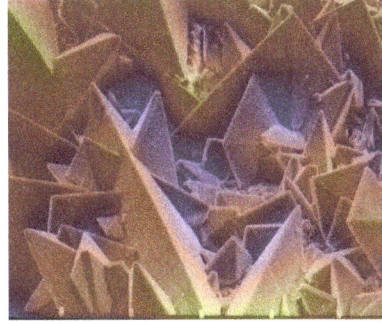

The same after re-processing of the color from the estimated topography

SEM image of a diagenetically altered discoaster

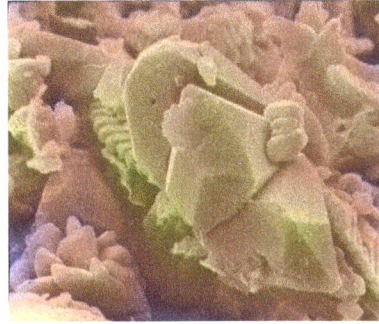

The same image after similar colorization

SEM Image Coloring

Very often, published SEM images are artificially colored. This may be done for aesthetic effect, to clarify structure or to add a realistic appearance to the sample and generally does not add information about the specimen.

Coloring may be performed manually with photo-editing software, or semi-automatically with dedicated software using feature-detection or object-oriented segmentation.

SEM image of *Cobaea scandens* pollen	The same after semi-automatic coloring. Arbitrary colors help identifying the various elements of the structure	Colored SEM image of *Tradescantia* pollen and stamens

Color Built using Multiple Electron Detectors

In some configurations more information is gathered per pixel, often by the use of multiple detectors.

As a common example, secondary electron and backscattered electron detectors are superimposed and a color is assigned to each of the images captured by each detector, with an end result of a combined color image where colors are related to the density of the components. This method is known as density-dependent color SEM (DDC-SEM). Micrographs produced by DDC-SEM retain topographical information, which is better captured by the secondary electrons detector and combine it to the information about density, obtained by the backscattered electron detector.

DDC-SEM of calcified particle in cardiac tissue - Signal 1 : SE	Signal 2 : BSE
Colorized image obtained from the two previous. Density-dependent color scanning electron micrograph SEM (DDC-SEM) of cardiovascular calcification, showing in orange a calcium phosphate spherical particle (denser material) and, in green, the extracellular matrix (less dense material)	Same work with a larger view, part of a study on human cardiovascular tissue calcification

Analytical Signals based on Generated Photons

Measurement of the energy of photons emitted from the specimen is a common method to get analytical capabilities. Examples are the Energy-dispersive X-ray spectroscopy (EDS) detectors used in elemental analysis and Cathodoluminescence microscope (CL) systems that analyse the intensity and spectrum of electron-induced luminescence in (for example) geological specimens. In SEM systems using these detectors it is common to color code these extra signals and superimpose them in a single color image, so that differences in the distribution of the various components of the specimen can be seen clearly and compared. Optionally, the standard secondary electron image can be merged with the one or more compositional channels, so that the specimen's structure and composition can be compared. Such images can be made while maintaining the full integrity of the original signal data, which is not modified in any way.

3D in SEM

SEMs do not naturally provide 3D images contrary to SPMs. However 3D data can be obtained using an SEM with different methods as follows:

3D SEM Reconstruction from a stereo pair

- photogrammetry (2 or 3 images from tilted specimen)

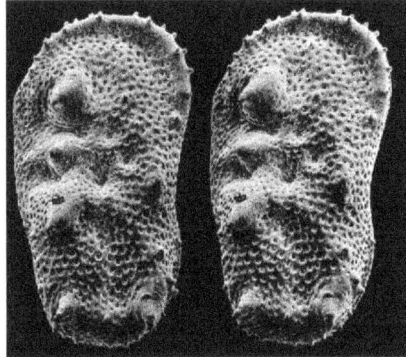

An SEM stereo pair of microfossils of less than 1 mm in size (Ostracoda) produced
by tilting along the longitudinal axis.

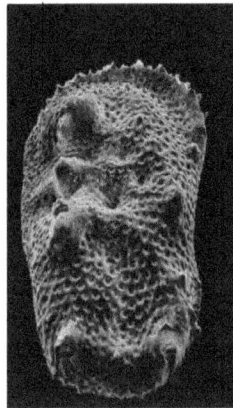

From this pair of SEM images, the third dimension has been reconstructed by photogrammetry
(using MountainsMap software) ; then a series of 3D representations with different angles
have been made.

Photometric 3D SEM Reconstruction from a Four-quadrant Detector "Shape from Shading"

This method typically uses a four-quadrant BSE detector. The microscope produces four images of the same specimen at the same time, so no tilt is required. The method gives metrological 3D dimensions as far as the slope of the specimen remains reasonable.

Some scanning electron microscopes are provided with software which uses a vendor-specific and usually closed-source algorithm to determine the 3D profile of the sample from the four-quadrant BSE detector. The supplied algorithms can be very simple line-by-line approaches, that only compare pixels next to each other. Due to the sample and conditions changing, this produces reasonable results only along the scan direction and is only practical for 1D line cuts along the scanning axis. Other approaches use more sophisticated (and sometimes GPU-intensive) methods like the optimal estimation algorithm and offer much better results at the cost of high demands on computing power.

As this approach works by integration of the slope, vertical slopes and overhangs are ignored; for instance, if an entire sphere lies on a flat, little more than the upper hemisphere is seen emerging above the flat, resulting in wrong altitude of the sphere apex. The prominence of this effect depends on the angle of the BSE detectors with respect to the sample, but these detectors are usually situated around (and close to) the electron beam, so this effect is very common.

Photometric 3D Rendering from a Single SEM Image

This method requires an SEM image obtained in oblique low angle lighting. The grey-level is then interpreted as the slope, and the slope integrated to restore the specimen topography. This method is interesting for visual enhancement and the detection of the shape and position of objects ; however the vertical heights cannot usually be calibrated, contrary to other methods such as photogrammetry.

SEM image of a house fly compound eye surface at 450× magnification.

Detail of the previous image.

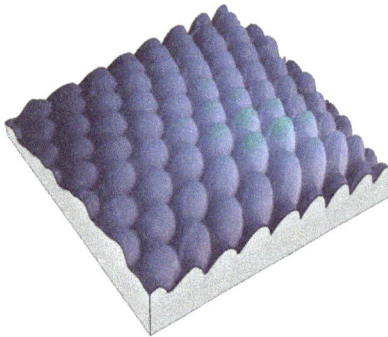

SEM 3D reconstruction from the previous using shape from shading algorithms.

Same as the previous, but with lighting homogenized before applying the shape from shading algorithms

Other Types of 3D SEM Reconstruction

- inverse reconstruction using electron-material interactive models.

- vertical stacks of SEM micrographs plus image-processing software.

- Multi-Resolution reconstruction using single 2D File: High-quality 3D imaging may be an ultimate solution for revealing the complexities of any porous media, but acquiring them is costly and time consuming. High-quality 2D SEM images, on the other hand, are widely available. Recently, a novel three-step, multiscale, multiresolution reconstruction method is presented that directly uses 2D images in order to develop 3D models. This method, based on a Shannon Entropy and conditional simulation, can be used for most of the available stationary materials and can build various stochastic 3D models just using a few thin sections.

Applications of 3D SEM

One possible application is measuring the roughness of ice crystals. This method can combine variable-pressure enviromental SEM and the 3D capabilites of the SEM to measure roughness on individual ice crystal facets, convert it into a computer model and run further statistical analysis on the model. Other measurements include fractal dimension, examining fracture surface of metals, characterization of materials, corrosion measurement, and dimensional measurements at the nano scale (step height, volume, angle, flatness, bearing ratio, coplanarity, etc.).

Apart from low resolving power one of the major limitations of optical microscope is its poor depth of focus. Therefore for examination of microstructure at higher magnification it is absolutely necessary to have perfectly flat surface. Likewise in TEM the total area which is investigated is extremely small. Therefore even though it gives vital information about the finer structural details it fails to give an overall idea of structural variations on a larger scale. These problems are overcome by adopting an entirely different way of accumulating structural details in Scanning Electron Microscope. It has extremely high depth of focus primarily because of the way the image is formed and magnified. The quality of the structure primarily depends on the ability of the unit to focus electron beam to as small an area as possible, the ability of the scanning coil to move this on the sample and the quality of the detectors to collect and amplify the signal.

Unlike TEM a much larger area can be examined under SEM within a very short time. It also allows examination of rough surfaces. Resolving power is also extremely high. It can reveal very fine structural details. One of the main advantages of SEM is that it does not need elaborate sample preparation procedure as in TEM. Rough fractured surface of metals and alloys can directly be examined. Normal specimens suitable for optical microscope are good enough for SEM. Modern SEM can also have detectors (EDS: Energy Dispersive Spectrometer) to pick up secondary emissions taking place from selected area of the surface and analyze the same to give local compositions. This is extremely useful in the identification of phases in microstructures. SEM can also have facility (as it scans the sample surface) to pick up back scattered diffracted beams of electrons. Analysis of the pattern can give the orientation of the grains. The data stored can be used to generate grain orientation maps and help identify characters of the grain boundaries. Such units are often called Orientation Imaging Microscope (OIM).

Tensile Testing

Tensile testing, is also known as tension testing, is a fundamental materials science test in which a sample is subjected to a controlled tension until failure. The results from the test are commonly used to select a material for an application, for quality control, and to predict how a material will react under other types of forces. Properties that are directly measured via a tensile test are ultimate tensile strength, maximum elongation and reduction in area. From these measurements the following properties can also be determined: Young's modulus, Poisson's ratio, yield strength, and strain-hardening characteristics. Uniaxial tensile testing is the most commonly used for obtaining the mechanical characteristics of isotropic materials. For anisotropic materials, such as composite materials and textiles, biaxial tensile testing is required.

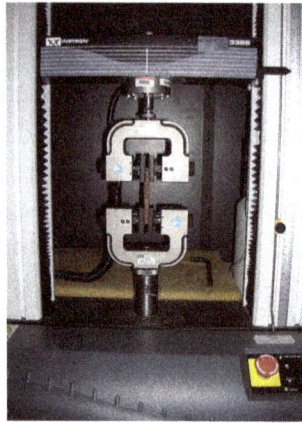

Tensile testing on a coir composite.

Tensile specimen

A tensile specimen is a standardized sample cross-section. It has two shoulders and a gage (section) in between. The shoulders are large so they can be readily gripped, whereas the gauge section has a smaller cross-section so that the deformation and failure can occur in this area.

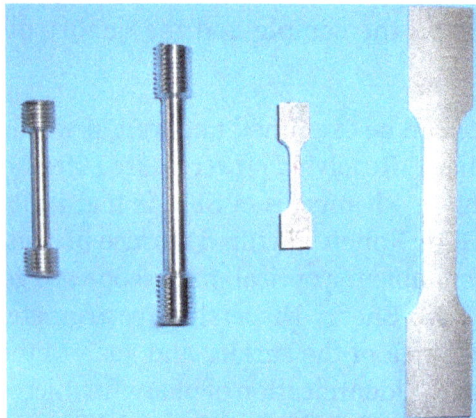

Tensile specimens made from an aluminum alloy. The left two specimens have a round cross-section and threaded shoulders. The right two are flat specimens designed to be used with serrated grips.

The shoulders of the test specimen can be manufactured in various ways to mate to various grips in the testing machine. Each system has advantages and disadvantages; for example, shoulders designed for serrated grips are easy and cheap to manufacture, but the alignment of the specimen is dependent on the skill of the technician. On the other hand, a pinned grip assures good alignment. Threaded shoulders and grips also assure good alignment, but the technician must know to thread each shoulder into the grip at least one diameter's length, otherwise the threads can strip before the specimen fractures.

In large castings and forgings it is common to add extra material, which is designed to be removed from the casting so that test specimens can be made from it. These specimens may not be exact representation of the whole workpiece because the grain structure may be different throughout.

In smaller workpieces or when critical parts of the casting must be tested, a workpiece may be sacrificed to make the test specimens. For workpieces that are machined from bar stock, the test specimen can be made from the same piece as the bar stock.

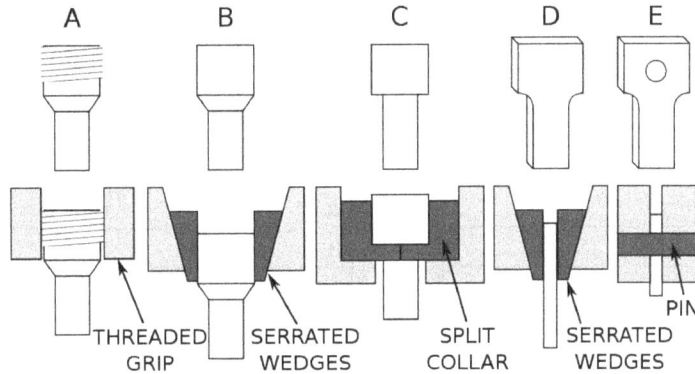

Various shoulder styles for tensile specimens. Keys A through C are for round specimens, whereas keys D and E are for flat specimens. Key:

A. A Threaded shoulder for use with a threaded grip

B. A round shoulder for use with serrated grips

C. A butt end shoulder for use with a split collar

D. A flat shoulder for used with serrated grips

E. A flat shoulder with a through hole for a pinned grip

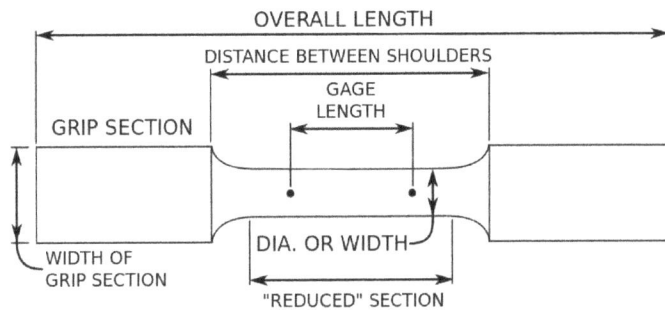

Test specimen nomenclature

The repeatability of a testing machine can be found by using special test specimens meticulously made to be as similar as possible.

A standard specimen is prepared in a round or a square section along the gauge length, depending on the standard used. Both ends of the specimens should have sufficient length and a surface condition such that they are firmly gripped during testing. The initial gauge length Lo is standardized (in several countries) and varies with the diameter (Do) or the cross-sectional area (Ao) of the specimen as listed

Type specimen	United States(ASTM)	Britain	Germany
Sheet (Lo / √Ao)	4.5	5.65	11.3
Rod (Lo / Do)	4.0	5.00	10.0

The following tables gives examples of test specimen dimensions and tolerances per standard ASTM E8.

Flat test specimen			
All values in inches	**Plate type (1.5 in. wide)**	**Sheet type (0.5 in. wide)**	**Sub-size specimen (0.25 in. wide)**
Gauge length	8.00±0.01	2.00±0.005	1.000±0.003
Width	1.5 +0.125−0.25	0.500±0.010	0.250±0.005
Thickness	$0.188 \leq T$	$0.005 \leq T \leq 0.75$	$0.005 \leq T \leq 0.25$
Fillet radius (min.)	1	0.25	0.25
Overall length (min.)	18	8	4
Length of reduced section (min.)	9	2.25	1.25
Length of grip section (min.)	3	2	1.25
Width of grip section (approx.)	2	0.75	$\frac{3}{8}$

Round test specimen					
All values in inches	**Standard specimen at nominal diameter:**		**Small specimen at nominal diameter:**		
	0.500	**0.350**	**0.25**	**0.160**	**0.113**
Gauge length	2.00±0.005	1.400±0.005	1.000±0.005	0.640±0.005	0.450±0.005
Diameter tolerance	±0.010	±0.007	±0.005	±0.003	±0.002
Fillet radius (min.)	$\frac{3}{8}$	0.25	$\frac{5}{16}$	$\frac{5}{32}$	$\frac{3}{32}$
Length of reduced section (min.)	2.5	1.75	1.25	0.75	$\frac{5}{8}$

Equipment

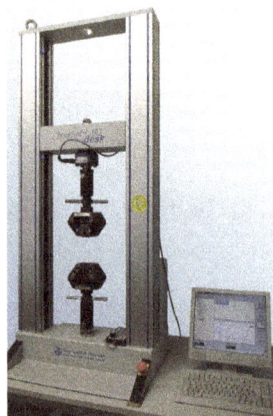

A universal testing machine (Hegewald & Peschke)

The most common testing machine used in tensile testing is the *universal testing machine*. This type of machine has two *crossheads*; one is adjusted for the length of the specimen and the other is driven to apply tension to the test specimen. There are two types: hydraulic powered and electromagnetically powered machines.

The machine must have the proper capabilities for the test specimen being tested. There are four main parameters: force capacity, speed, precision and accuracy. Force capacity refers to the fact that the machine must be able to generate enough force to fracture the specimen. The machine must be able to apply the force quickly or slowly enough to properly mimic the actual application. Finally, the machine must be able to accurately and precisely measure the gauge length and forces applied; for instance, a large machine that is designed to measure long elongations may not work with a brittle material that experiences short elongations prior to fracturing.

Alignment of the test specimen in the testing machine is critical, because if the specimen is misaligned, either at an angle or offset to one side, the machine will exert a bending force on the specimen. This is especially bad for brittle materials, because it will dramatically skew the results. This situation can be minimized by using spherical seats or U-joints between the grips and the test machine. If the initial portion of the stress–strain curve is curved and not linear, it indicates the specimen is misaligned in the testing machine.

The strain measurements are most commonly measured with an extensometer, but strain gauges are also frequently used on small test specimen or when Poisson's ratio is being measured. Newer test machines have digital time, force, and elongation measurement systems consisting of electronic sensors connected to a data collection device (often a computer) and software to manipulate and output the data. However, analog machines continue to meet and exceed ASTM, NIST, and ASM metal tensile testing accuracy requirements, continuing to be used today.

Process

The test process involves placing the test specimen in the testing machine and slowly extending it until it fractures. During this process, the elongation of the gauge section is recorded against the applied force. The data is manipulated so that it is not specific to the geometry of the test sample. The elongation measurement is used to calculate the *engineering strain*, ε, using the following equation:

$$\varepsilon = \frac{\Delta L}{L_0} = \frac{L - L_0}{L_0}$$

where ΔL is the change in gauge length, L_0 is the initial gauge length, and L is the final length. The force measurement is used to calculate the *engineering stress*, σ, using the following equation:

$$\sigma = \frac{F_n}{A}$$

where F is the tensile force and A is the nominal cross-section of the specimen. The machine does these calculations as the force increases, so that the data points can be graphed into a *stress–strain curve*.

Standards

Metals

- ASTM E8/E8M-13: "Standard Test Methods for Tension Testing of Metallic Materials" (2013).

- ISO 6892-1: "Metallic materials. Tensile testing. Method of test at ambient temperature" (2009).

- ISO 6892-2: "Metallic materials. Tensile testing. Method of test at elevated temperature" (2011).

- JIS Z2241 Method of tensile test for metallic materials.

Flexible Materials

- ASTM D638 Standard Test Method for Tensile Properties of Plastics.

- ASTM D828 Standard test method for tensile properties of paper and paperboard using constant-rate-of-elongation apparatus.

- ASTM D882 Standard test method for tensile properties of thin plastic sheeting.

- ISO 37 rubber, vulcanized or thermoplastic—determination of tensile stress–strain properties.

A specimen of a standard size and shape made of the material is pulled in tension while the load and elongation are continuously (or periodically) monitored. The data thus collected are used to generate stress strain curve. Modern machines have an intelligent computer interface and have facility to perform tests under various user defined modes (for example load control, displacement control and strain control). Usually tensile tests are performed under displacement control mode where the moveable end of the grip is made to move at a constant speed. The load and displacement (strain) records are stored at specified intervals of time. Once the sample breaks the machine stops. The data can then be displayed on the computer screen.

Table: Main characteristics of elastic and plastic deformation

Elastic deformation		Reversible
		Stress strain relation: Hooke's law: $CJ\ EE$
		Metals mostly exhibit linear elastic behaviour
Plastic deformation		Irreversible: leaves permanent strain
		Stress strain beyond YS: $CJ\ KE^n$ K & n are constants. Slope of unloading step is same as that of the elastic part of the plot.

For most engineering applications the magnitude of tensile stress is defined as the load over the initial cross sectional area. However as strain increases the area decreases therefore the true stress at any given strain should be defined as the load over the instantaneous area. As long as the magnitude of strain is small the difference between the two is insignificant. When we talk of large strain well within the plastic regime it is more appropriate to consider true stress and strain (defined as the increase in the gauge length over the instantaneous gauge length).

$$\text{True strain} = \varepsilon = \int_{L_0}^{L} \frac{dL}{L} = ln\left(\frac{L}{L_0}\right) = ln\left(\frac{L_0 + \Delta L}{L_0}\right) = ln(1+e) \qquad (3)$$

$$\text{True stress} = \sigma = \frac{P}{A} = \frac{P}{A_0}\left(\frac{A_0}{P}\right) = S\left(\frac{A_0}{A}\right) = S\left(\frac{L_0}{L}\right) = S(1+e) \qquad (4)$$

Note that if the deformation is uniform and there is no change in volume due to plastic deformation $A_0 L_0 = AL$. Using equation 3 & 4, the engineering stress strain curves can be converted to true stress strain plots. The relation between stress & strain beyond YS given in table are valid for true stress and true strain.

The local deformation or necking as it is more commonly known as is encountered during tensile testing of ductile material. This is also known as tensile instability. Assuming a simple relationship between stress and plastic strain as given in table it is possible to derive the condition under which such instability is likely to occur. This is described below:

$$\text{Applied stress} = \sigma = \frac{P}{A} \qquad (5)$$

$$\text{It can be also written as}: \frac{d\sigma}{\sigma} = \frac{dP}{P} - \frac{dA}{A} \qquad (6)$$

When the load reaches its maximum value: $\frac{dP}{P} = 0. \left(-\frac{dA}{A}\right)$ still represents incremental strain.

Therefore it is possible to derive the condition at which necking sets in. This is given by:

$$\sigma = \frac{d\sigma}{d\varepsilon} \qquad (7)$$

Equation 7 states that at necking the slope of the stress strain plot becomes equal to the stress itself. The slope of the stress strain plot gives an idea about the extent of strain hardening. Since the relation between stress and strain is given by:

$$\sigma = K\varepsilon^n; \therefore \frac{d\sigma}{d\varepsilon} = Kn\varepsilon^{n-1} = n\frac{\sigma}{\varepsilon} \qquad (8)$$

from equation 7 & 8 it follows that $\varepsilon = n$ $\qquad$ (9)

This shows that n is an indicator of the strain at necking. A large n means the material has good ductility. For most metals it is around 0.3.

Table: Shows how various mechanical properties of a material can be obtained from its stress strain plot

Proportional limit (PL)		Indicates the stress up to which it is directly proportional to strain.

0.2% Proof stress (PS)		Sometimes it is difficult to identify the yield point. Therefore the stress to reach a specified level of strain (0.002) is taken as the yield stress.
Resilience (RS)		A measure of recoverable stored energy, given by the area under the elastic part of the stress strain diagram. Useful property for the selection of material for spring.
Toughness (T)		The area under the total stress strain plot minus the amount of recoverable energy due to elastic deformation. This is a measure of energy absorbed due to plastic deformation.

Hardening (Metallurgy)

Hardening is a metallurgical metalworking process used to increase the hardness of a metal. The hardness of a metal is directly proportional to the uniaxial yield stress at the location of the imposed strain. A harder metal will have a higher resistance to plastic deformation than a less hard metal.

Processes

The five hardening processes are:

- The Hall–Petch method, or grain boundary strengthening, is to obtain small grains. Smaller grains increases the likelihood of dislocations running into grain boundaries after shorter distances, which are very strong dislocation barriers. In general, smaller grain size will make the material harder. When the grain size approach sub-micron sizes, some materials may however become softer. This is simply an effect of another deformation mechanism that becomes easier, i.e. grain boundary sliding. At this point, all dislocation related hardening mechanisms become irrelevant.

- In work hardening (also referred to as strain hardening or cold working) the material is strained past its yield point. The plastic straining generate new dislocations. As the dislocation density increases, further dislocation movement becomes more difficult since they hinder each other, which means the material hardness increases.

- In solid solution strengthening, a soluble alloying element is added to the material desired to be strengthened, and together they form a "solid solution". A solid solution can be thought of just as a "normal" liquid solution, e.g. salt in water, except it is solid. Depending on the size of the dissolved alloying element's ion compared to that of the matrix-metal, it is dissolved either substitutionally (large alloying element substituting for an atom in the crystal) or interstitially (small alloying element taking a place between atoms in the crystal lattice). In both cases, the size difference of the foreign elements make them act as sand grains in sandpaper, resisting dislocations that try to slip by, resulting in higher material strength. In solution hardening, the alloying element does not precipitate from solution.

- Precipitation hardening (also called *age hardening*) is a process where a second phase that begins in solid solution with the matrix metal is precipitated out of solution with the metal as it is quenched, leaving particles of that phase distributed throughout to cause resistance to slip dislocations. This is achieved by first heating the metal to a temperature where the elements forming the particles are soluble then quenching it, trapping them in a solid solution. Had it been a liquid solution, the elements would form precipitates, just as supersaturated saltwater would precipitate small salt crystals, but atom diffusion in a solid is very slow at room temperature. A second heat treatment at a suitable temperature is then required to age the material. The elevated temperature allows the dissolved elements to diffuse much faster, and form the desired precipitated particles. The quenching is required since the material otherwise would start the precipitation already during the slow cooling. This type of precipitation results in few large particles rather than the, generally desired, profusion of small precipitates. Precipitation hardening is one of the most commonly used techniques for the hardening of metal alloys.

- Martensitic transformation, more commonly known as quenching and tempering, is a hardening mechanism specific for steel. The steel must be heated to a temperature where the iron phase changes from ferrite into austenite, i.e. changes crystal structure from BCC (body-centered cubic) to FCC (face-centered cubic). In austenitic form, steel can dissolve a lot more carbon. Once the carbon has been dissolved, the material is then quenched. It is important to quench with a high cooling rate so that the carbon does not have time to form precipitates of carbides. When the temperature is low enough, the steel tries to return to the low temperature crystal structure BCC. This change is very quick since it does not rely on diffusion and is called a martensitic transformation. Because of the extreme supersaturation of solid solution carbon, the crystal lattice becomes BCT (body-centered tetragonal) instead. This phase is called martensite, and is extremely hard due to a combined effect of the distorted crystal structure and the extreme solid solution strengthening, both mechanisms of which resist slip dislocation.

All hardening mechanisms introduce crystal lattice defects that act as barriers to dislocation slip.

Applications

Material hardening is required for many applications:

- Construction materials - High strength reduces the need for material thickness which generally saves weight and cost.

- Machine cutting tools (drill bits, taps, lathe tools) need be much harder than the material they are operating on in order to be effective.

- Knife blades – a high hardness blade keeps a sharp edge.

- Bearings – necessary to have a very hard surface that will withstand continued stresses.

- Armor plating - High strength is extremely important both for bullet proof plates and for heavy duty containers for mining and construction.

- Anti-fatigue - (Martensitic) case hardening can drastically improve the service life of mechanical components with repeated loading/unloading, such as axles and cogs.

Indentation Hardness

Indentation hardness tests are used in mechanical engineering to determine the hardness of a material to deformation. Several such tests exist, wherein the examined material is indented until an impression is formed; these tests can be performed on a macroscopic or microscopic scale.

When testing metals, indentation hardness correlates roughly linearly with tensile strength., but it is an imperfect correlation often limited to small ranges of strength and hardness for each indentation geometry. This relation permits economically important nondestructive testing of bulk metal deliveries with lightweight, even portable equipment, such as hand-held Rockwell hardness testers.

Material Hardness

As of the direction of materials science continues towards studying the basis of properties on smaller and smaller scales, different techniques are used to quantify material characteristics and tendencies. Measuring mechanical properties for materials on smaller scales, like thin films, can not be done using conventional uniaxial tensile testing. As a result, techniques testing material "hardness" by indenting a material with an impression have been developed to determine such properties.

Hardness measurements quantify the resistance of a material to plastic deformation. Indentation hardness tests compose the majority of processes used to determine material hardness, and can be divided into two classes: *microindentation* and *macroindentation* tests. Microindentation tests typically have forces less than 2 N (0.45 lb_f). Hardness, however, cannot be considered to be a fundamental material property. Instead, it represents an arbitrary quantity used to provide a relative idea of material properties. As such, hardness can only offer a comparative idea of the material's resistance to plastic deformation since different hardness techniques have different scales.

The main sources of error with indentation tests are poor technique, poor calibration of the equipment, and the strain hardening effect of the process. However, it has been experimentally determined through "strainless hardness tests" that the effect is minimal with smaller indentations.

Surface finish of the part and the indenter do not have an effect on the hardness measurement, as long as the indentation is large compared to the surface roughness. This proves to be useful when measuring the hardness of practical surfaces. It also is helpful when leaving a shallow indentation, because a finely etched indenter leaves a much easier to read indentation than a smooth indenter.

The indentation that is left after the indenter and load are removed is known to "recover", or spring back slightly. This effect is properly known as *shallowing*. For spherical indenters the indentation is known to stay symmetrical and spherical, but with a larger radius. For very hard materials the radius can be three times as large as the indenter's radius. This effect is attributed to the release of elastic stresses. Because of this effect the diameter and depth of the indentation do contain errors. The error from the change in diameter is known to be only a few percent, with the error for the depth being greater.

Another effect the load has on the indentation is the *piling-up* or *sinking-in* of the surrounding material. If the metal is work hardened it has a tendency to pile up and form a "crater". If the metal is annealed it will sink in around the indentation. Both of these effects add to the error of the hardness measurement.

The equation based definition of hardness is the pressure applied over the contact area between the indenter and the material being tested. As a result hardness values are typically reported in units of pressure, although this is only a "true" pressure if the indenter and surface interface is perfectly flat.

Macroindentation Tests

The term "macroindentation" is applied to tests with a larger test load, such as 1 kgf or more. There are various macroindentation tests, including:

- Vickers hardness test (HV), which has one of the widest scales

- Brinell hardness test (HB)

- Knoop hardness test (HK), for measurement over small areas

- Janka hardness test, for wood

- Meyer hardness test

- Rockwell hardness test (HR), principally used in the USA

- Shore hardness test, for polymers

- Barcol hardness test, for composite materials.

There is, in general, no simple relationship between the results of different hardness tests. Though there are practical conversion tables for hard steels, for example, some materials show qualitatively different behaviors under the various measurement methods. The Vickers and Brinell hardness scales correlate well over a wide range, however, with Brinell only producing overestimated values at high loads.

Microindentation Tests

The term "microhardness" has been widely employed in the literature to describe the hardness testing of materials with low applied loads. A more precise term is "microindentation hardness testing." In microindentation hardness testing, a diamond indenter of specific geometry is impressed into the surface of the test specimen using a known applied force (commonly called a "load" or "test load") of 1 to 1000 gf. Microindentation tests typically have forces of 2 N (roughly 200 gf) and produce indentations of about 50 μm. Due to their specificity, microhardness testing can be used to observe changes in hardness on the microscopic scale. Unfortunately, it is difficult to standardize microhardness measurements; it has been found that the microhardness of almost any material is higher than its macrohardness. Additionally, microhardness values vary with load and work-hardening effects of materials. The two most commonly used microhardness tests are tests that also can be applied with heavier loads as macroindentation tests:

- Vickers hardness test (HV)

- Knoop hardness test (HK)

In microindentation testing, the hardness number is based on measurements made of the indent formed in the surface of the test specimen. The hardness number is based on the applied force divided by the surface area of the indent itself, giving hardness units in kgf/mm². Microindentation hardness testing can be done using Vickers as well as Knoop indenters. For the Vickers test, both the diagonals are measured and the average value is used to compute the Vickers pyramid number. In the Knoop test, only the longer diagonal is measured, and the Knoop hardness is calculated based on the projected area of the indent divided by the applied force, also giving test units in kgf/mm².

The Vickers microindentation test is carried out in a similar manner welling to the Vickers macroindentation tests, using the same pyramid. The Knoop test uses an elongated pyramid to indent material samples. This elongated pyramid creates a shallow impression, which is beneficial for measuring the hardness of brittle materials or thin components. Both the Knoop and Vickers indenters require prepolishing of the surface to achieve accurate results.

Scratch tests at low loads, such as the Bierbaum microcharacter test, performed with either 3 gf or 9 gf loads, preceded the development of microhardness testers using traditional indenters. In 1925, Smith and Sandland of the UK developed an indentation test that employed a square-based pyramidal indenter made from diamond. They chose the pyramidal shape with an angle of 136° between opposite faces in order to obtain hardness numbers that would be as close as possible to Brinell hardness numbers for the specimen. The Vickers test has a great advantage of using one hardness scale to test all materials. The first reference to the Vickers indenter with low loads was made in the annual report of the National Physical Laboratory in 1932. Lips and Sack describes the first Vickers tester using low loads in 1936.

There is some disagreement in the literature regarding the load range applicable to microhardness testing. ASTM Specification E384, for example, states that the load range for microhardness testing is 1 to 1000 gf. For loads of 1 kgf and below, the Vickers hardness (HV) is calculated with an equation, wherein load (L) is in grams force and the mean of two diagonals (d) is in millimeters:

$$HV = 0.0018544 \times \frac{L}{d^2}$$

For any given load, the hardness increases rapidly at low diagonal lengths, with the effect becoming more pronounced as the load decreases. Thus at low loads, small measurement errors will produce large hardness deviations. Thus one should always use the highest possible load in any test. Also, in the vertical portion of the curves, small measurement errors will produce large hardness deviations.

Hardness

It is often defined as the ability of a material to resist scratch or indentation. Talc is known for its softness. You may be able to scratch it with your nail. In terms of Moh's scale (Oldest measure of hardness) hardness of talc is taken as zero and that of diamond (the strongest solid) as 10. All materials have hardness values in between these limits. However this is too qualitative a measure to be of practical significance for engineering application. Resistance to deformation or indentation is a more practical and popular measure of hardness. There are several different ways of estimating this by trying to force an indenter made of a hard non- deformable material of standard dimension. The hardness is defined on the basis of the type of indenter, load used and the size of the impression left on the material. Three most commonly used indentation hardness scales are known as Rockwell, Brinell and Vickers. Rockwell hardness is a linear function of depth of indentation. Higher the depth of indentation lower is the hardness. The scale ranges between 0-100. It uses either a diamond 120° cone indenter or a ball indenter made of hardened steel. Depending on the combination of indenter and load there are several Rockwell hardness scales. Three most commonly used Rockwell hardness scales are given in table below. The tester consists of a moveable sample stand, a loading device connected to an indenter. The sample is placed beneath an indenter. A minor load of 10kg is applied. The position of the indenter is indicated on a dial gauge fixed on the loading device. The reading at this stage corresponds to zero. Subsequently the major load is applied. The indenter penetrates into the material. After a specified time interval the major load is withdrawn. The indenter moves up but does not come back to its original position. The scale is calibrated in a reverse fashion. The closer it comes back to its original position higher is the hardness. The combination of indenter and load are so chosen that the reading lies between 20 and 80. Rc scale is used to measure hardness of heat treated steel whereas RB scale is used for relatively softer materials like aluminium.

Table: Common Rockwell Hardness Scale

Rockwell Scale	Minor load, kg	Major load, kg	Total load, kg	Indenter
R_A	10	50	60	120° diamond cone
R_B	10	90	100	1/16" steel ball
R_C	10	140	150	120° diamond cone

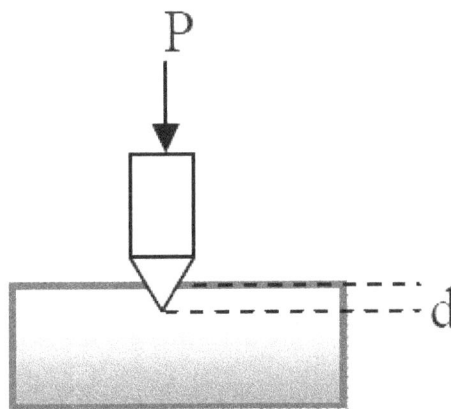

The extent of penetration by major load P in a Rockwell hardness tester.
It measures d after the major load is withdrawn. H = K-d

Brinell hardness uses either 10 or 5mm diameter ball indenter made of either hardened steel or tungsten carbide. The applied load depends on the hardness of material. As a thumb rule the load used for measuring the hardness of steel – $30D^a$ kg; where D is the diameter of the ball. If D = 10mm the load to be used = 3000kg. The machine consists of a loading device connected to the indenter and a moveable stand to place the sample. The sample stand is raised so that it touches the indenter. Thereafter the specified load is applied. After a specified length of hold time (~10s) the load is withdrawn. This leaves an indentation mark. Its diameter (d) is measured using a graduated eyepiece. If P is the applied load the Brinell hardness number (BHN) of the material is given by load / area of the indentation. Unlike Rockwell hardness which is just a number BHN has a dimension of kg/mm2. The expression to estimate BHN is as follows:

$$BHN= \frac{2P}{\pi D(D - \sqrt{D^2 - d^2})} \qquad (10)$$

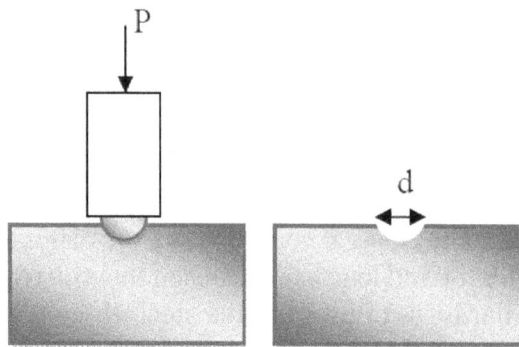

The extent of penetration by load P in a Brinell hardness tester.
It measures diameter (d) of the indentation using a graduated eye piece after the load is withdrawn.
Hardness is given by P divided by the surface area of the indentation.

Since the loads and the sizes of indenter are much larger than those used for Rockwell scale the impression left on the sample too is much larger. It is likely to give an average hardness over a larger area of the sample. BHN is a preferred scale if the material is heterogeneous (example cast iron). Vickers hardness scale (number) is similar to that of BHN. The difference lies in the choice of indenter and the load. It uses square based diamond pyramid indenter with an apex angle of 136°. The loads used are in the range 1 to 120kg. This scale is independent of load. Since the load is much lower than that in Brinell, the size of indentation is much smaller. It needs a microscope to measure the size of the indentation mark. The commercial hardness measurement system is equipped with a loading device connected to the indenter; an adjustable specimen stage and a microscope with provision to measure the diagonal of the impression left behind or an imaging system than can magnify the indentation mark to facilitate measurement. Hardness is defined as load over the area of the indentation. Vickers Hardness Number (VHN) is given by:

$$VHN = \frac{2P}{d^2}Sin\left(\frac{\theta}{2}\right) = 1.8534\frac{P}{d^2} \qquad (11)$$

Both BHN & VHN use the similar concept for the measurement of hardness. Therefore for most materials the hardness values are nearly the same up to 500BHN (approximately). Since indentation hardness measures the resistance to deformation it has a direct correlation with tensile strength of the material. The hardness number is approximately 3 times the UTS in MPa. Measurement of hardness is extremely simple and easy. It does not need elaborate sample preparation or machining. A flat surface is good enough. Only VHN needs a little better surface finish with fine emery paper. Hardness measurements using these three scales are very popular for engineering applications. It (Rockwell & VHN) is nearly a non-destructive method of estimating the strength of a material. Whatever technique you use for hardness measurement there are standard conversion tables to help you convert hardness measured on one scale to another.

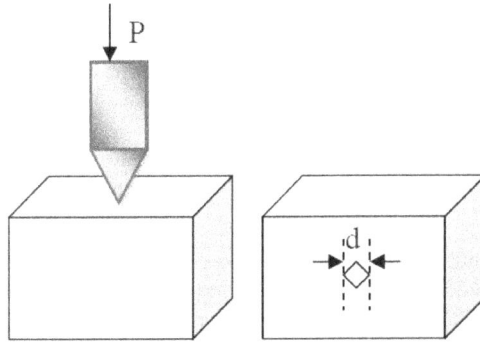

The extent of penetration by load P in a Vickers hardness tester.
It measures diagonal (d) of the indentation using a graduated eye piece after the load is withdrawn.
Hardness is given by P divided by the surface area of the indentation.

Impact

Often materials known to be ductile and tough under normal rate of loading is found be brittle if it is subjected to impact or sudden loading. Impact testing helps us assess the ability of a material to withstand such loading. There are several methods of testing impact resistance. However we shall try to give an idea about one of the most common techniques known as Charpy V Notch (CVN) test. This uses a standard 10mm square 55mm long specimen with a 2mm deep V notch located half way between the two ends. It is placed horizontally on two supports at either ends so that the V notch is vertical. A hammer having a specified shape and size mounted on fixture that can be made to swing like a pendulum is used to strike the specimen on the face opposite the one having the V notch. The hammer can be raised to different heights as necessary. The height gives the net potential energy stored in the hammer (= mgh; where m is the mass of the hammer in kg, h is the height in meter (m) and g is acceleration due to gravity in m/s²). When it is released this gets converted into kinetic energy. When the hammer is in its lowest position, its kinetic energy or its velocity attains its maximum value $\left(= \frac{1}{2}mv^2 \right)$. This is when it strikes the sample. As a result the sample breaks and the hammer still continues to swing. The height to which it rises immediately after it breaks the sample gives an estimate of the energy of the hammer after the sample is broken. It breaks through initiation and propagation of a crack from the notch root. The difference between the two gives a measure of the energy absorbed by the sample. Higher the amount of energy needed to break the specimen higher is its impact resistance. It is expressed in terms of Joule. When it absorbs a large amount of energy to break there are signs of large local deformation near the notch root. The fractured surface is rough. Such materials are said to be tough. A brittle material on the other hand absorbs little energy for crack initiation and propagation. There is hardly any sign of notch root deformation. The fractured surface is nearly flat. CVN is found to be a function of materials, its microstructure and the test temperature. Metals by and large have CVN. However steel is known to exhibit transition from ductile behaviour at room (and high temperature) to brittle behaviour at low temperature. Figure below shows a typical CVN versus temperature plot of steel. The temperature at which there is a sharp change in CVN value is known as its transition temperature.

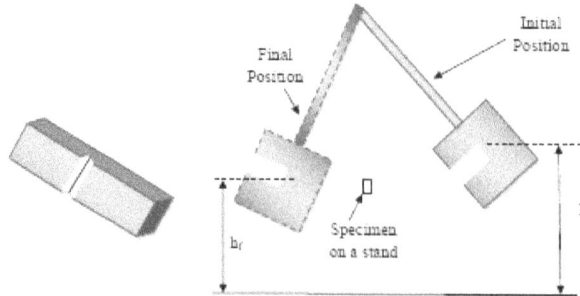

a) The standard CVN test piece (55mm long 10X10mm cross section) with a 2mm wide v shaped notch on one face as shown. The hammer strikes the face opposite the one having the notch. b) The position of hammer just before and after the test. The hammer can swing like a pendulum. When released from a height of hi it goes to other side to the same height if there is no specimen. If there is a specimen on the stand as shown it absorbs a part of the energy of the hammer. Let the height the hammer rises to now be hf. Therefore CVN = mg (hi-hf).

Limitation of CVN as a measure of toughness: It only gives a qualitative index of the ability of a material to resist unstable growth of a notch that represents a defect under impact loading. Unlike properties such as yield strength it does not allow estimation of the load a component could withstand from the CVN data. This is because in the presence of a defect the local state of stress near the defect is more complex than that in the case of a uniformly stressed sample. The stress that helps the crack to open up or propagate is perpendicular to the plane on which it lies. Figure below represents the orientation of a crack of length equal to a, in a sample supported at two ends and loaded at the centre. This is commonly known as 3 point bend loading. This is the way a sample is loaded during Charpy V-notch impact testing. It also gives the nature of the stress distribution near the crack tip. The concept of K1C which is known as the fracture toughness of the material has also been explained.

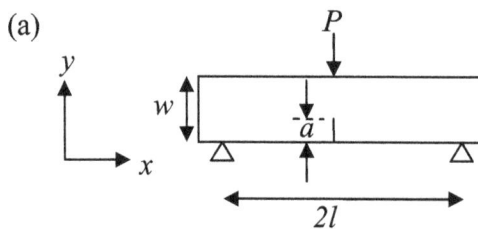

(c)
$$\sigma_x = \frac{K_1}{\sqrt{2\pi y}} F(x, y)$$

$$K_1 = \sigma\sqrt{\pi a}\, f(a/w)$$

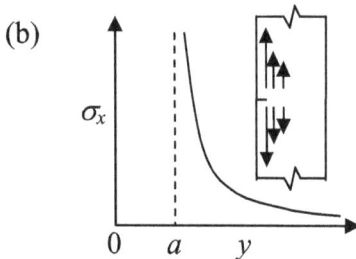

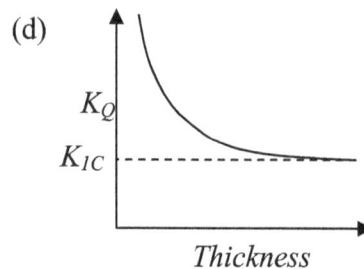

(a) Shows sample of width w, thickness B and length $2l$ is loaded in bending. The load P is applied at the centre of the span. The sample has a crack of length a as shown.

(b) Shows how the stress acting along x axis varies with the distance from the crack tip (y).

(c) Gives the dependence of stress σ_x on stress intensity factor (SIF) K_1 and a geometric function F. SIF gives the amplitude of the stress distribution near the crack tip. The expression for K_1 has a term $f(a/w)$. This depends on the loading and geometry. Unstable crack extension occurs as it approaches K1C representing the fracture toughness of the material.

(d) Shows the effect of sample thickness on the magnitude of the SIF at which unstable crack growth occurs in a specimen of thickness B.

For simplicity let us assume $f(a/w) = 1$. This gives the following expression for the stress at which a crack of length 'a' would grow in an unstable manner.

$$\sigma_f = \frac{K_{1C}}{\sqrt{\pi a}} \qquad (12)$$

Equation 4 suggests that in the presence of a crack the stress at which fracture takes palce could be substantially low. Expressions for $f(a/w)$ various types of specimen and loadings are available in any books on mechanical behavior of engineering materials or fracture mechanics

Fatigue

When a material is subjected to cyclic loading it fails after a certain number of cycles of loading even if the load much lower than its yields strength. This phenomenon is known as fatigue. Many engineering components are subjected to cyclic loading (for example any rotating shaft, connecting rod, rails, wheels etc). The resistance of a material to withstand such a failure is a function of loading and material characteristic.

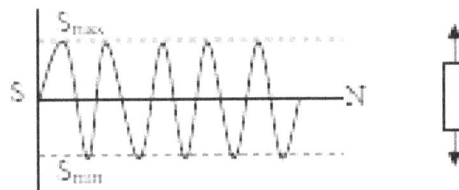

A diagram showing a sample being subjected to cyclic loading. The stress keeps changing between two limits Smax & Smin. The stress amplitude S = (Smax-Smin) / 2. The test continues till failure. Such tests are carried out at different stress amplitude to generate S versus N (number of cycles to failure) plot. This is known as S-N plot.

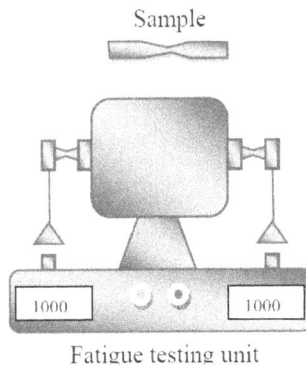

Fatigue testing unit

The sketch of a rotating beam fatigue testing unit with two samples mounted on it with the loading train.

An enlarged view of test specimen made from cylindrical rod is shown just above the testing unit. The sample is loaded as a cantilever. Therefore it has a reduced section at the centre to ensure that the central portion of the test piece is subjected to maximum stress (bending moment). Two samples can be mounted on the machine. Each has separate loading pan fixed with the help of bearing. The top half of the sample is subjected to tension whereas the bottom half is subjected to compression. As the sample rotates the stress at point keeps changing. This is how it is subjected to cyclic loading.

As the sample rotates the stress at a point keeps changing. One rotation corresponds to one cycle of loading. The machine has a counter to count & display the number of cycles of loading. When the sample breaks the loading pan falls on a limit switch that stops the machine and the counter. The number of cycles to failure is noted. Tests are conducted at different stress ranges. The stress (S: the peak stress when the mean load is zero) versus number of cycles to failure plot is popularly known as S-N curve. Such data help in estimating fatigue lives of engineering components subjected to cyclic loading. For materials like steel the S-N curve slope becomes zero below a specific value of stress S. This is called the endurance limit. Within this stress range the component is expected to have infinite life. However there are a host of other metals & alloys that do not show a definite endurance limit. These are the materials that do not have infinite fatigue life at any stress range.

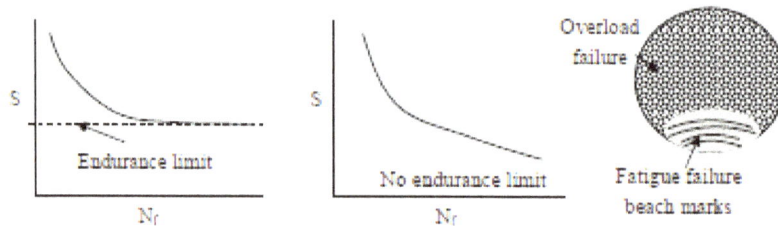

A typical plot for fatigue

a) A typical S-N curve for steel showing a definite endurance limit. It is a plot of stress amplitude versus the number of cycles to failure (2Nf). Below the endurance limit the sample can withstand infinite numbers of cycle.

b) A typical S-N curve for Al alloys. It has no definite endurance limit. In such cases 10^7 cycles is taken as the limit.

c) A typical appearance of fatigue fracture the curved lines denote beach marks. Fatigue failure takes place by initiation and propagation of crack. Once a crack develops the two faces that come in contact during subsequent stages of loading get rubbed by each other. This results in beach marks. When the remaining area becomes too small to support the load rapid fracture takes place. This may have features of normal tensile failure. The fractured surface thus has two distinct regions one having beach marks characteristic of fatigue loading and the other relatively rough face typical of overload failure.

Beach mark as shown in figure above is the characteristic signature of fatigue failure. It suggests that under cyclic loading failure does not take place all on a sudden. It has two distinct stages: nucleation and growth. A tiny crack first nucleates (often at the exposed surface of the specimen or a component) and later it propagates leaving behind such beach marks. Examination of the plastic replica taken from the beach marked region of a fractured surface under TEM (Transmission

Electron Microscope) suggests that these are made of several striation marks consisting of a set of parallel lines. The distance between two consecutive striations denotes crack extension during each cycle of loading.

S-N curves are usually obtained from fatigue tests performed on rotating beam fatigue testing units. The stress over the entire cross section of the specimen is not uniform. Once a crack initiates the peak stress at the crack tip too increases significantly. This would result in an ever increasing crack growth rate. Therefore it is more likely to give an estimate of fatigue crack initiation life. The S-N curve can be represented by Basquin equation. It is given by

$$\sigma_a = \sigma_f (2N_f)^b \tag{13}$$

It means if a failure takes place during the first half cycle when the load is tensile the stress amplitude corresponds to the fracture stress of the material. The magnitude of b is often in the range of 0.1 – 0.2 for most metals and alloys. It gives a conservative estimate of the fatigue life of an engineering material.

Over the last 5 decades there has been a significant progress in our understanding of the fatigue crack growth behaviour. This has been facilitated by the availability of servo-hydraulic and servoelectric material testing system where standard samples having pre-existing cracks can be subjected to user defined uniaxial fatigue loading. In the presence of a crack the stress at the tip of a crack is best described by stress intensity factor (K). The relation between fatigue crack growth $\left(\dfrac{da}{dN}\right)$ and stress intensity range (ΔK) is best described by the following equation which is commonly known as Paris law.

$$\frac{da}{dN} = c(\Delta K)^n \tag{14}$$

In this expression c and n are material constant. The magnitude of n is 4 for several commercial structural materials.

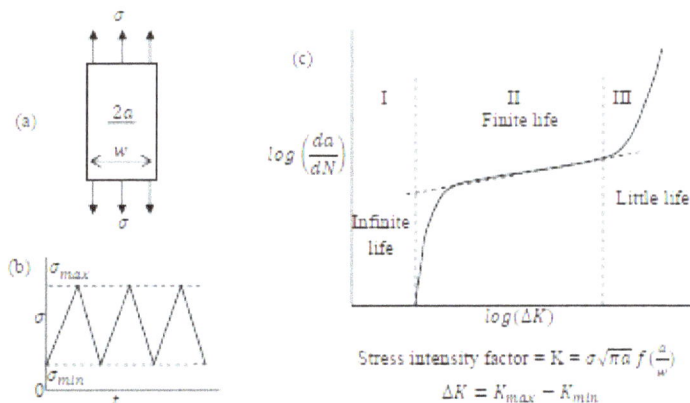

A schematic representation of the results obtained from a test performed on a centre cracked tension panel having a crack of length 2a at its centre.

(a) Shows a test specimen of width w having a crack of length $2a$ at its centre. Such specimens are commonly known as centre cracked tension panel. It is subjected to uniform tensile stress σ.

(b) Shows how the stress keep changing with time between two limits σ_{max} and σ_{min}.

(c) Shows how the rate of crack extension per cycle of loading increases with an increase in the stress intensity range (ΔK). It has 3 distinct stages. The stage I has infinite life. In stage II the crack growth rate can be expressed as a function of ΔK. In stage III the crack growth rate keeps increasing rapidly.

The comparison of the two methods of estimating fatigue life suggests that the endurance limit in the case of the S-N curve approach corresponds to the first stage of the fatigue crack growth behaviour. It depends on the microstructure of a material. It also gives a measure of the size of the defect that does not grow under a given fatigue loading.

Creep (Deformation)

In materials science, creep (sometimes called cold flow) is the tendency of a solid material to move slowly or deform permanently under the influence of mechanical stresses. It can occur as a result of long-term exposure to high levels of stress that are still below the yield strength of the material. Creep is more severe in materials that are subjected to heat for long periods, and generally increases as they near their melting point.

The rate of deformation is a function of the material properties, exposure time, exposure temperature and the applied structural load. Depending on the magnitude of the applied stress and its duration, the deformation may become so large that a component can no longer perform its function — for example creep of a turbine blade will cause the blade to contact the casing, resulting in the failure of the blade. Creep is usually of concern to engineers and metallurgists when evaluating components that operate under high stresses or high temperatures. Creep is a deformation mechanism that may or may not constitute a failure mode. For example, moderate creep in concrete is sometimes welcomed because it relieves tensile stresses that might otherwise lead to cracking.

Unlike brittle fracture, creep deformation does not occur suddenly upon the application of stress. Instead, strain accumulates as a result of long-term stress. Therefore, creep is a "time-dependent" deformation.

Temperature Dependence

The temperature range in which creep deformation may occur differs in various materials. For example, tungsten requires a temperature in the thousands of degrees before creep deformation can occur, while ice will creep at temperatures near 0 °C (32 °F). As a general guideline, the effects of creep deformation generally become noticeable at approximately 35% of the melting point (as measured on a thermodynamic temperature scale such as Kelvin or Rankine) for metals, and at 45% of melting point for ceramics. Virtually any material will creep upon approaching its melting temperature. Since the creep minimum temperature is related to the melting point, creep can be seen at relatively low temperatures for some materials. Plastics and low-melting-temperature metals, including many solders, can begin to creep at room temperature, as can be seen markedly in old lead hot-water pipes. Glacier flow is an example of creep processes in ice.

Stages of Creep

In the initial stage, or primary creep, or transient creep, the strain rate is relatively high, but decreases with increasing time and strain due to a process analogous to work hardening at lower temperatures. For instance, the dislocation density increases and, in many materials, a dislocation subgrain structure is formed and the cell size decreases with strain. The strain rate diminishes to a minimum and becomes near constant as the secondary stage begins. This is due to the balance between work hardening and annealing (thermal softening). The secondary stage, referred to as "steady-state creep", is the most understood. The microstructure is invariant during this stage, which means that recovery effects are concurrent with deformation. No material strength is lost during these first two stages of creep.

The characterized "creep strain rate" typically refers to the constant rate in this secondary stage. Stress dependence of this rate depends on the creep mechanism. In tertiary creep, the strain rate exponentially increases with stress because of necking phenomena or internal cracks or voids decreases the effective area of the specimen. Strength is quickly lost in this stage while the material's shape is permanently changed. The acceleration of creep deformation in the tertiary stage eventually leads to material fracture.

Mechanisms of Creep

The mechanism of creep depends on temperature and stress. Under the conditions of different temperature and applied stress, dislocation glide, dislocation climb, or diffusional-flow mechanisms may dominate creep deformation. Some mechanisms of creep, especially those involving dislocations, have not been verified by direct microstructural examination yet. However, processes just like the mechanisms conjectured should happen during creep deformation.

Various mechanisms are:

- Bulk diffusion (Nabarro-Herring creep)
- Climb — here the strain is actually accomplished by climb
- Climb-assisted glide — here the climb is an *enabling* mechanism, allowing dislocations to get around obstacles
- Grain boundary diffusion (Coble creep)
- Thermally activated glide — e.g., via cross-slip

General Creep Equation

$$\frac{d\varepsilon}{dt} = \frac{C\sigma^m}{d^b} e^{\frac{-Q}{kT}}$$

where ε is the creep strain, C is a constant dependent on the material and the particular creep mechanism, m and b are exponents dependent on the creep mechanism, Q is the activation energy of the creep mechanism, σ is the applied stress, d is the grain size of the material, k is Boltzmann's constant, and T is the absolute temperature.

Dislocation Creep

At high stresses (relative to the shear modulus), creep is controlled by the movement of dislocations. For dislocation creep, $Q = Q$(self diffusion), $m = 4–6$, and $b = 0$. Therefore, dislocation creep
has a strong dependence on the applied stress and the intrinsic activation energy, but no grain size
dependence.

Some alloys exhibit a very large stress exponent ($n > 10$), and this has typically been explained by
introducing a "threshold stress," σ_{th}, below which creep can't be measured. The modified power
law equation then becomes:

$$\frac{d\varepsilon}{dt} = A\left(\sigma - \sigma_{th}\right)^n e^{\frac{-Q}{RT}}$$

where A, Q and n can all be explained by conventional mechanisms (so $3 \le n \le 10$). The creep
increases with increasing applied stress, since the applied stress tends to drive the dislocation
past the barrier, and make the dislocation get into a lower energy state after bypassing the obstacle, which means that the dislocation is inclined to pass the obstacle. In other words, part of the
work required to overcome the energy barrier of passing an obstacle is provided by the applied
stress and the remainder by thermal energy.

Nabarro–Herring Creep

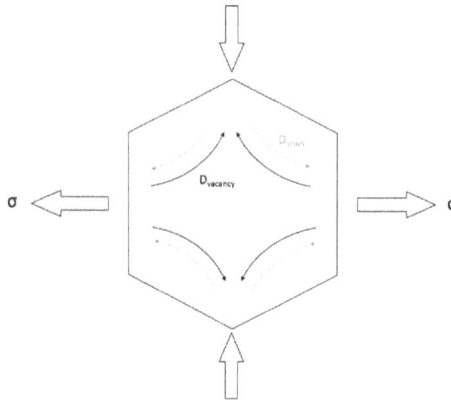

A diagram showing the diffusion of atoms and vacancies under Nabarro–Herring Creep.

Nabarro–Herring (NH) creep is a form of diffusion creep, while dislocation glide creep does not
involve atomic diffusion. Nabarro-Herring creep dominates at high temperatures and low stresses.
As shown in the figure, the lateral sides of the crystal are subjected to a tensile stress, and the horizontal sides to a compressive stress. The atomic volume is altered by applied stress: it increases
in regions under tension and decreases in regions under compression. So the activation energy for
vacancy formation is changed by $\pm\sigma\Omega$ where Ω is the atomic volume, the "$+$" sign is for compressive regions and "$-$" sign is for tensile regions. Since the fractional vacancy concentration is
proportional to $\exp(-\frac{Q_f \pm \sigma\Omega}{RT})$, where Q_f is the vacancy-formation energy, the vacancy concentration is higher in tensile regions than in compressive regions, leading to a net flow of vacancies
from the regions under tension to the regions under compression, and this is equivalent to a net

atom diffusion in the opposite direction, which causes the creep deformation: the grain elongates in the tensile stress axis and contracts in the compressive stress axis.

In Nabarro–Herring creep, k is related to the diffusion coefficient of atoms through the lattice, $Q = Q$ (self diffusion), $m = 1$, and $b = 2$. Therefore, Nabarro–Herring creep has a weak stress dependence and a moderate grain size dependence, with the creep rate decreasing as grain size is increased.

Nabarro–Herring creep is strongly temperature dependent. For lattice diffusion of atoms to occur in a material, neighboring lattice sites or interstitial sites in the crystal structure must be free. A given atom must also overcome the energy barrier to move from its current site (it lies in an energetically favorable potential well) to the nearby vacant site (another potential well). The general form of the diffusion equation is $D = D_o \exp(E/KT)$ where D_o has a dependence on both the attempted jump frequency and the number of nearest neighbor sites and the probability of the sites being vacant. Thus there is a double dependence upon temperature. At higher temperatures the diffusivity increases due to the direct temperature dependence of the equation, the increase in vacancies through Schottky defect formation, and an increase in the average energy of atoms in the material. Nabarro–Herring creep dominates at very high temperatures relative to a material's melting temperature.

Coble Creep

Coble creep is a second form of diffusion controlled creep. In Coble creep the atoms diffuse along grain boundaries to elongate the grains along the stress axis. This causes Coble creep to have a stronger grain size dependence than Nabarro–Herring creep, thus, Coble creep will be more important in materials composed of very fine grains. For Coble creep k is related to the diffusion coefficient of atoms along the grain boundary, $Q = Q$(grain boundary diffusion), $m = 1$, and $b = 3$. Because Q(grain boundary diffusion) < Q(self diffusion), Coble creep occurs at lower temperatures than Nabarro–Herring creep. Coble creep is still temperature dependent, as the temperature increases so does the grain boundary diffusion. However, since the number of nearest neighbors is effectively limited along the interface of the grains, and thermal generation of vacancies along the boundaries is less prevalent, the temperature dependence is not as strong as in Nabarro–Herring creep. It also exhibits the same linear dependence on stress as Nabarro–Herring creep. Generally, the diffusional creep rate should be the sum of Nabarro–Herring creep rate and Coble creep rate. Diffusional creep leads to grain-boundary separation, that is, voids or cracks form between the grains. To heal this, grain-boundary sliding occurs. The diffusional creep rate and the grain boundary sliding rate must be balanced if there are no voids or cracks remain. When grain-boundary sliding couldn't accommodate the incompatibility, grain-boundary voids are generated, which is related to the initiation of creep fracture.

Solute Drag Creep

Solute drag creep is one kind of mechanisms for power law creep (PLC), involving both dislocation and diffusional flow. Solute drag creep is observed in certain metallic alloys. Their creep rate increases during the first stage of creep before a steady-state, which can be explained by a model associated with solid-solution strengthening. The size misfit between solute atoms and edge dislocations could restrict dislocation motion. The flow stress required for dislocations to move is increased at low temperatures due to immobility of the solute atoms. But solute atoms are mobile at higher temperatures, so the solute atoms could move along with edge dislocations as a "drag" on their motion, if the dislocation motion or the creep rate is not too high. The solute drag creep rate is:

$$\frac{d\varepsilon}{dt} = C\frac{D_{sol}\sigma^3}{\varepsilon_b^2 c_0}$$

where C is a constant, D_{sol} is the solute diffusivity, c_0 is the solute concentration, and ε_b is the misfit parameter, σ is the applied stress. So it could be seen from the equation above, m is 3 for solute drag creep. Solute drag creep shows a special phenomenon, which is called the Portevin-Le Chatelier effect. When the applied stress becomes sufficiently large, the dislocations will break away from the solute atoms since dislocation velocity increases with the stress. After breakaway, the stress decreases and the dislocation velocity also decreases, which allows the solute atoms to approach and reach the previously departed dislocations again, leading to a stress increase. The process repeats itself when the next local stress maximum is obtained. So repetitive local stress maxima and minima could be detected during solute drag creep.

Dislocation Climb-glide Creep

Dislocation climb-glide creep is observed in materials at high temperature. The initial creep rate is larger than the steady-state creep rate. Climb-glide creep could be illustrated as follows: when the applied stress is not enough to for a moving dislocation to overcome the obstacle on its way via dislocation glide alone, the dislocation could climb to a parallel slip plane by diffusional processes, and the dislocation can glide on the new plane. This process repeats itself each time when the dislocation encounters an obstacle. The creep rate could be written as:

$$\frac{d\varepsilon}{dt} = \frac{A_{CG}D_L}{M^{1/2}}(\frac{\sigma\Omega}{kT})^{4.5}$$

where A_{CG} includes details of the dislocation loop geometry, D_L is the lattice diffusivity, M is the number of dislocation sources per unit volume, σ is the applied stress, and Ω is the atomic volume. The exponent m for dislocation climb-glide creep is 4.5 if M is independent of stress and this value of m is consistent with results from considerable experimental studies.

Harper–Dorn Creep

Harper–Dorn creep is a climb-controlled dislocation mechanism at low stresses that has been observed in aluminum, lead, and tin systems, in addition to nonmetal systems such as ceramics and ice. It is characterized by two principal phenomena: a linear relationship between the steady-state strain rate and applied stress at a constant temperature, and an independent relationship between the steady-state strain rate and grain size for a provided temperature and applied stress. The latter observation implies that Harper–Dorn creep is controlled by dislocation movement; namely, since creep can occur by vacancy diffusion (Nabarro–Herring creep, Coble creep), grain boundary sliding, and/or dislocation movement, and since the first two mechanisms are grain-size dependent, Harper–Dorn creep must therefore be dislocation-motion dependent.

However, Harper–Dorn creep is typically overwhelmed by other creep mechanisms in most situations, and is therefore not observed in most systems. The phenomenological equation which describes Harper–Dorn creep is:

$$\frac{d\varepsilon}{dt} = \rho_0 \frac{D_v Gb^3}{k_B T}\left(\frac{\sigma_s}{G}\right)$$

where: ρ_0 is dislocation density (constant for Harper–Dorn creep), D_v is the diffusivity through the volume of the material, G is the shear modulus, b is the Burger's vector, σ_s is the applied stress, k_B is Boltzmann's constant, and T is temperature.

The volumetric activation energy indicates that the rate of Harper–Dorn creep is controlled by vacancy diffusion to and from dislocations, resulting in climb-controlled dislocation motion. Unlike in other creep mechanisms, the dislocation density here is constant and independent of the applied stress. Moreover, the dislocation density must be low for Harper–Dorn creep to dominate. The density has been proposed to increase as dislocations move via cross-slip from one slip-plane to another, thereby increasing the dislocation length per unit volume. Cross-slip can also result in jogs along the length of the dislocation, which, if large enough, can act as single-ended dislocation sources.

Examples

Creep of Polymers

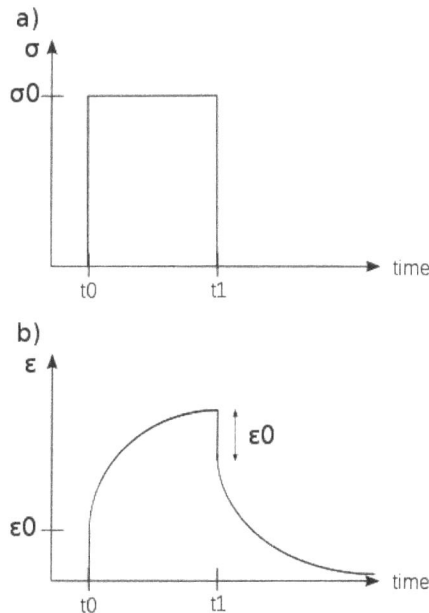

a) Applied stress and b) induced strain as functions of time over a short period for a viscoelastic material.

Creep can occur in polymers and metals which are considered viscoelastic materials. When a polymeric material is subjected to an abrupt force, the response can be modeled using the Kelvin-Voigt model. In this model, the material is represented by a Hookean spring and a Newtonian dashpot in parallel. The creep strain is given by the following convolution integral:

$$\varepsilon(t) = \sigma C_0 + \sigma C \int_0^\infty f(\tau)(1 - \exp[-t/\tau])d\tau$$

where:

σ = applied stress

C_0 = instantaneous creep compliance

C = creep compliance coefficient

τ = retardation time

$f(\tau)$ = distribution of retardation times

When subjected to a step constant stress, viscoelastic materials experience a time-dependent increase in strain. This phenomenon is known as viscoelastic creep.

At a time t_0, a viscoelastic material is loaded with a constant stress that is maintained for a sufficiently long time period. The material responds to the stress with a strain that increases until the material ultimately fails. When the stress is maintained for a shorter time period, the material undergoes an initial strain until a time t_1 at which the stress is relieved, at which time the strain immediately decreases (discontinuity) then continues decreasing gradually to a residual strain.

Viscoelastic creep data can be presented in one of two ways. Total strain can be plotted as a function of time for a given temperature or temperatures. Below a critical value of applied stress, a material may exhibit linear viscoelasticity. Above this critical stress, the creep rate grows disproportionately faster. The second way of graphically presenting viscoelastic creep in a material is by plotting the creep modulus (constant applied stress divided by total strain at a particular time) as a function of time. Below its critical stress, the viscoelastic creep modulus is independent of stress applied. A family of curves describing strain versus time response to various applied stress may be represented by a single viscoelastic creep modulus versus time curve if the applied stresses are below the material's critical stress value.

Additionally, the molecular weight of the polymer of interest is known to affect its creep behavior. The effect of increasing molecular weight tends to promote secondary bonding between polymer chains and thus make the polymer more creep resistant. Similarly, aromatic polymers are even more creep resistant due to the added stiffness from the rings. Both molecular weight and aromatic rings add to polymers' thermal stability, increasing the creep resistance of a polymer.

Both polymers and metals can creep. Polymers experience significant creep at temperatures above ca. −200 °C; however, there are three main differences between polymeric and metallic creep.

Polymers show creep basically in two different ways. At typical work loads (5 up to 50%) ultra high molecular weight polyethylene (Spectra, Dyneema) will show time-linear creep, whereas polyester or aramids (Twaron, Kevlar) will show a time-logarithmic creep.

Creep of Concrete

The creep of concrete, which originates from the calcium silicate hydrates (C-S-H) in the hardened

Portland cement paste (which is the binder of mineral aggregates), is fundamentally different from the creep of metals as well as polymers. Unlike the creep of metals, it occurs at all stress levels and, within the service stress range, is linearly dependent on the stress if the pore water content is constant. Unlike the creep of polymers and metals, it exhibits multi-months aging, caused by chemical hardening due to hydration which stiffens the microstructure, and multi-year aging, caused by long-term relaxation of self-equilibrated micro-stresses in the nano-porous microstructure of the C-S-H. If concrete is fully dried it does not creep, though it is difficult to dry concrete fully without severe cracking.

Applications

Creep on the underside of a cardboard box: a largely empty box was placed on a smaller box, and more boxes were placed on top of it. Due to the weight, the portions of the empty box not upheld by the lower support gradually deflected downward.

Though mostly due to the reduced yield strength at higher temperatures, the collapse of the World Trade Center was due in part to creep from increased temperature operation.

The creep rate of hot pressure-loaded components in a nuclear reactor at power can be a significant design constraint, since the creep rate is enhanced by the flux of energetic particles.

Creep in epoxy anchor adhesive was blamed for the Big Dig tunnel ceiling collapse in Boston, Massachusetts that occurred in July 2006.

The design of tungsten light bulb filaments attempts to reduce creep deformation. Sagging of the filament coil between its supports increases with time due to the weight of the filament itself. If too much deformation occurs, the adjacent turns of the coil touch one another, causing an electrical short and local overheating, which quickly leads to failure of the filament. The coil geometry and supports are therefore designed to limit the stresses caused by the weight of the filament, and a special tungsten alloy with small amounts of oxygen trapped in the crystallite grain boundaries is used to slow the rate of Coble creep.

Creep can cause gradual cut-through of wire insulation, especially when stress is concentrated by pressing insulated wire against a sharp edge or corner. Special creep-resistant insulations such as Kynar (polyvinylidene fluoride) are used in wirewrap applications to resist cut-through due to the sharp corners of wire wrap terminals. Teflon insulation is resistant to elevated temperatures and has other desirable properties, but is notoriously vulnerable to cold-flow cut-through failures caused by creep.

In steam turbine power plants, pipes carry steam at high temperatures (566 °C (1,051 °F)) and pressures (above 24.1 MPa or 3500 psi). In jet engines, temperatures can reach up to 1,400 °C (2,550 °F) and initiate creep deformation in even advanced-design coated turbine blades. Hence, it is crucial for correct functionality to understand the creep deformation behavior of materials.

Creep deformation is important not only in systems where high temperatures are endured such as nuclear power plants, jet engines and heat exchangers, but also in the design of many everyday objects. For example, metal paper clips are stronger than plastic ones because plastics creep at room temperatures. Aging glass windows are often erroneously used as an example of this phenomenon: measurable creep would only occur at temperatures above the glass transition temperature around 500 °C (932 °F). While glass does exhibit creep under the right conditions, apparent sagging in old windows may instead be a consequence of obsolete manufacturing processes, such as that used to create crown glass, which resulted in inconsistent thickness.

Fractal geometry, using a deterministic Cantor structure, is used to model the surface topography, where recent advancements in thermoviscoelastic creep contact of rough surfaces are introduced. Various viscoelastic idealizations are used to model the surface materials, for example, Maxwell, Kelvin-Voigt, Standard Linear Solid and Jeffrey media.

Nimonic 75 has been certified by the European Union as a standard creep reference material.

Preventing Creep

There are three general ways to prevent creep in metal. One way is to use higher melting temperature metals. The second way is to use materials with greater grain size. The third way is to use alloying.

Creep of Superalloys

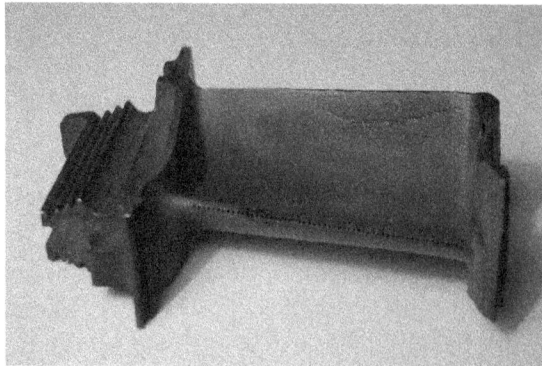

Materials operating at high temperatures, such as this nickel superalloy jet engine (RB199) turbine blade, must be able to withstand the significant creep present at these temperatures.

Materials operating in high-performance systems, such as jet engines, often reach extreme temperatures surpassing 1200 °C, which causes creep to be a serious issue. Superalloys based on Co, Ni, and Fe are capable of being engineered to be highly resistant to creep, and have thus arisen as an ideal material in high-temperature environments. As an example, Ni-base alloys modeled after the Ni-Al system, known as γ-γ' alloys are even resistant to dislocation creep. The γ is the main fcc matrix, while the γ' is the precipitate-phase of $Ni_3(Al, Ti)$, which adds particle strengthening.

Solute elements added, e.g., Ta, W, Mo, Fe, Cr, and Co, contribute solid-solution hardening, and are often reacted with carbon to form carbide particles that deposit at grain boundaries, and thus inhibit grain boundary sliding.

When a material is subjected to load at an elevated temperature it undergoes a slow time dependent deformation even if the stress is much lower than its yield strength. This phenomenon is known as creep. This is measurable at temperatures T°K > 0.4 TM°K where TM denotes melting point in °Kelvin. It has a loading device connected to a set of pull rods with grips to hold the sample, a furnace having a sufficiently long heating zone with temperature controller, and a strain measuring system. The sample is kept under constant load at a constant temperature and the strain is measured as a function of time. This has three distinct stages apart from the instantaneous strain on loading. Stage I where the strain rate keeps decreasing, stage II where strain rate is nearly constant and the stage III where the strain rate keeps increasing with time.

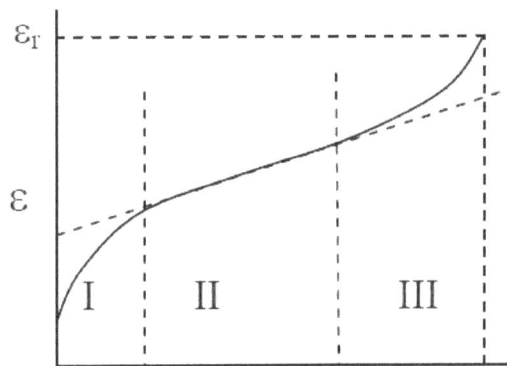

A typical strain time plot at a constant temperature and load.

This is known as creep curve. It has three stages: I called primary where strain rate keeps decreasing, II called steady (secondary) state and III called tertiary stage where creep rate keeps increasing with time. Time at which rupture takes place (t_r) & strain (ε_r) at rupture, steady state creep rate $\dot{\varepsilon}_s$ are some of the important parameters obtained from the test. These are functions of load (initial stress) and temperature. Since load remains constant as strain increases with time stress would continue to increase.

Creep behaviour of engineering materials is a strong function of stress, temperature and the internal structure of materials. Figure shows the effects of stress and temperature on the shape of the strain time plots. Creep strain accumulation increases significantly with stress and temperature. All high temperature components are subjected to creep. Therefore it becomes a major consideration for the selection of materials for such applications. There are a few simple relations derived from the analysis of experimental data that are generally applicable for most materials. A few of these are given below.

$$\text{Minimum (steady state creep rate)} = \dot{\varepsilon}_s = A \exp\left(-\frac{Q}{RT}\right)\sigma^n \qquad (15)$$

$$\dot{\varepsilon}_s t_r = \text{Monkman Grant constant} \qquad (16)$$

Note that $\dot{\varepsilon}_s$ & R in equation 13 denote steady state creep rate, and universal gas constant, Q = activation energy and A and n are material parameters (constants). In equation 16 tr denote rupture life (time to failure).

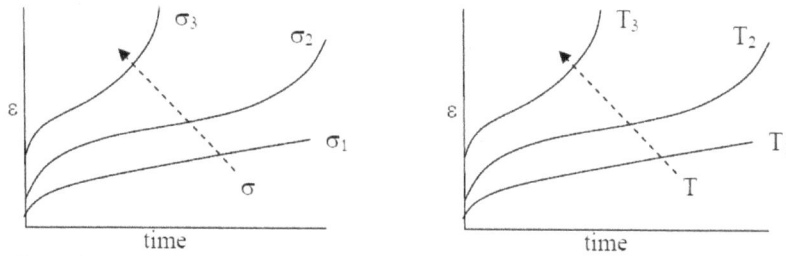

effect of stress b) effect of temperature on the shapes of creep strain time plots.

Clearly it appears that a large volume of data needs to be collected (some of the tests may continue for enormously long periods of time e.g., 10 years) and analysed for describing load bearing capacity of materials at elevated temperatures. This problem is overcome for most engineering applications by looking at the times to rupture (failure).

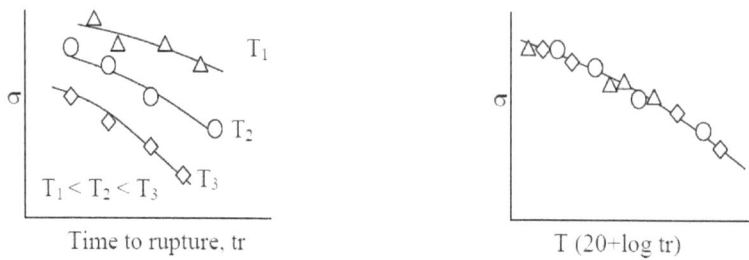

The effects of stress and temperature of time to rupture. Such curves are known as stress rupture plots.

(a) Shows rupture plots at three different temperatures. The lines show trend whereas symbols experimental points.

(b) If these data are plotted as the function of a combined time temperature parameter as shown one gets a common master rupture plot.

Note that irrespective of test temperature all data fall on the same trend line. The parameter used here {T(20+log tr), where T is in °Kelvin} is commonly known as Larson Miller parameter. There are a large numbers of similar parameters used in estimating stress rupture lives. The master rupture plot sometime may appear to have two distinct lines with different slopes. One at higher stresses has a relatively lower slope and the other at lower stress has a higher slope. The failure at higher stress level is due to trans-granular fracture whereas that at lower stress is due to inter-granular fracture.

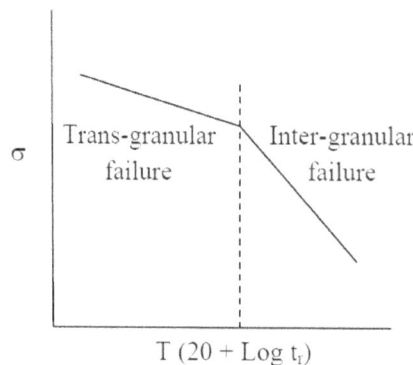

Stress rupture plots of most engineering materials may have two distinct regions. At higher stresses the failure is due to trans-granular fracture whereas at lower stresses failure is due to inter-granular fracture

References

- "Computer-aided cooling curve thermal analysis of near eutectic Al–Si–Cu–Fe alloy" (PDF). Journal of Thermal Analysis and Calorimetry. 114: 705–717. doi:10.1007/s10973-013-3005-7

- Albert Van Helden; Sven Dupré; Rob van Gent (2010). The Origins of the Telescope. Amsterdam University Press. pp. 32–36, 43. ISBN 978-90-6984-615-6

- Chang, Kenneth (8 October 2014). "2 Americans and a German Are Awarded Nobel Prize in Chemistry". New York Times. Retrieved 8 October 2014

- Paulik, F; Paulik, J; Erdey, L (1966). "Derivatography A complex method in thermal analysis". Talanta. 13: 1405–30. PMID 18960022. doi:10.1016/0039-9140(66)80083-8

- Manuel Gunkel; et al. (2009). "Dual color localization microscopy of cellular nanostructures". Biotechnology journal. 4 (6): 927–38. PMID 19548231. doi:10.1002/biot.200900005

- Gould, Stephen Jay (2000). "Chapter 2: The Sharp-Eyed Lynx, Outfoxed by Nature". The Lying Stones of Marrakech: Penultimate Reflections in Natural History. New York, N.Y: Harmony. ISBN 0-224-05044-3

- Kenneth, Spring; Keller, H. Ernst; Davidson, Michael W. "Microscope objectives". Olympus Microscopy Resource Center. Retrieved 29 Oct 2008

- Crewe, Albert V; Isaacson, M. and Johnson, D.; Johnson, D. (1969). "A Simple Scanning Electron Microscope". Rev. Sci. Inst. 40 (2): 241–246. Bibcode:1969RScI...40..241C. doi:10.1063/1.1683910

- Li, Z; Baker, ML; Jiang, W; Estes, MK; Prasad, BV (2009). "Rotavirus Architecture at Subnanometer Resolution". Journal of Virology. 83 (4): 1754–1766. PMC 2643745. PMID 19036817. doi:10.1128/JVI.01855-08

- Murphy, Douglas B. (2002). Fundamentals of Light Microscopy and Electronic Imaging. New York: John Wiley & Sons. ISBN 9780471234296

- "Ceiling Collapse in the Interstate 90 Connector Tunnel". National Transportation Safety Board. Washington, D.C.: NTSB. July 10, 2007. Retrieved 2 December 2016

- Crewe, Albert V; Wall, J. and Langmore, J., J; Langmore, J (1970). "Visibility of a single atom". Science. 168 (3937): 1338–1340. Bibcode:1970Sci...168.1338C. PMID 17731040. doi:10.1126/science.168.3937.1338

- Baram, M. & Kaplan W. D. (2008). "Quantitative HRTEM analysis of FIB prepared specimens". Journal of Microscopy. 232 (3): 395–05. PMID 19094016. doi:10.1111/j.1365-2818.2008.02134.x

- Stokes, Debbie J. (2008). Principles and Practice of Variable Pressure Environmental Scanning Electron Microscopy (VP-ESEM). Chichester: John Wiley & Sons. ISBN 978-0470758748

- "Next Monday, Digital Surf to Launch Revolutionary SEM Image Colorization". AZO Materials. Retrieved 23 January 2016

- McMullan, D. (1953). "An improved scanning electron microscope for opaque specimens". Proceedings of the IEE – Part II: Power Engineering. 100 (75): 245–256. doi:10.1049/pi-2.1953.0095

- Hindmarsh, J. P.; Russell, A. B.; Chen, X. D. (2007). "Fundamentals of the spray freezing of foods—microstructure of frozen droplets". Journal of Food Engineering. 78 (1): 136–150. doi:10.1016/j.jfoodeng.2005.09.011

- Ernst Ruska; translation by T Mulvey. The Early Development of Electron Lenses and Electron Microscopy. ISBN 3-7776-0364-3

- Nebesářová1, Jana; Vancová, Marie (2007). "How to Observe Small Biological Objects in Low-Voltage Electron Microscope". Microscopy and Microanalysis. 13 (3): 248–249. Retrieved 8 August 2016

- P.A. Crozier & T.W. Hansen (2014). "In situ and operando transmission electron microscopy of catalytic materials". MRS Bulletin. 40: 38–45. doi:10.1557/mrs.2014.304

Solidification and Deformation of Pure Metal

The chapter deals with solidification of metals, the factors pertaining to its form, and its transformation and affecting factors, and also to deformation. It also explores cooling curve, a graph that represents shift in the state of matter. This chapter has been carefully written to provide an easy understanding of the varied facets of solidification and deformation.

Cooling Curve

A cooling curve is a line graph that represents the change of phase of matter, typically from a gas to a solid or a liquid to a solid. The independent variable (X-axis) is time and the dependent variable (Y-axis) is temperature. Below is an example of a cooling curve used in castings.

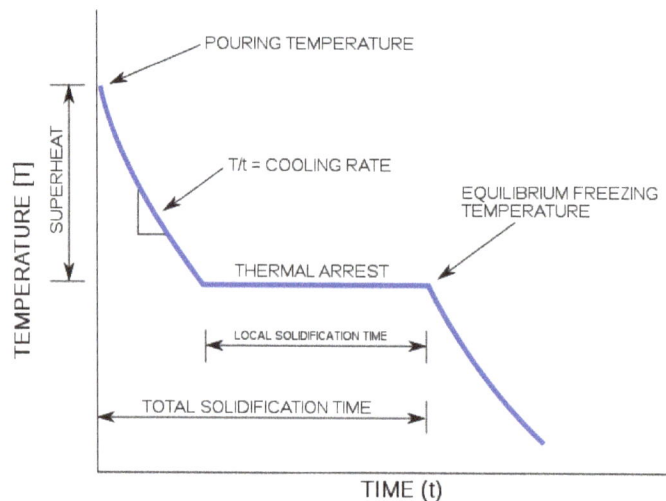

The initial point of the graph is the starting temperature of the matter, here noted as the "pouring temperature". When the phase change occurs there is a "thermal arrest", that is the temperature stays constant. This is because the matter has more internal energy as a liquid or gas than in the state that it is cooling to. The amount of energy required for a phase change is known as latent heat. The "cooling rate" is the slope of the cooling curve at any point.

Phase Rule

Gibbs's phase rule was proposed by Josiah Willard Gibbs in his landmark paper titled *On the Equilibrium of Heterogeneous Substances*, published from 1875 to 1878. The rule applies to non-reactive multi-component heterogeneous systems in thermodynamic equilibrium and is given by the equality

$$F = C - P + 2$$

where F is the number of degrees of freedom, C is the number of components and P is the number of phases in thermodynamic equilibrium with each other.

The number of degrees of freedom is the number of independent intensive variables, i.e. the largest number of thermodynamic parameters such as temperature or pressure that can be varied simultaneously and arbitrarily without affecting one another. An example of one-component system is a system involving one pure chemical, while two-component systems, such as mixtures of water and ethanol, have two chemically independent components, and so on. Typical phases are solids, liquids and gases.

Foundations

- A phase is a form of matter that is homogeneous in chemical composition and physical state. Typical phases are solid, liquid and gas. Two immiscible liquids (or liquid mixtures with different compositions) separated by a distinct boundary are counted as two different phases, as are two immiscible solids.

- The number of components (C) is the number of chemically independent constituents of the system, i.e. the minimum number of independent species necessary to define the composition of all phases of the system. For examples: component (thermodynamics).

- The number of degrees of freedom (F) in this context is the number of intensive variables which are independent of each other.

The basis for the rule (Atkins and de Paula, justification 6.1) is that equilibrium between phases places a constraint on the intensive variables. More rigorously, since the phases are in thermodynamic equilibrium with each other, the chemical potentials of the phases must be equal. The number of equality relationships determines the number of degrees of freedom. For example, if the chemical potentials of a liquid and of its vapour depend on temperature (T) and pressure (p), the equality of chemical potentials will mean that each of those variables will be dependent on the other. Mathematically, the equation $\mu_{liq}(T, p) = \mu_{vap}(T, p)$, where μ = chemical potential, defines temperature as a function of pressure or vice versa. (Caution: do not confuse p = pressure with P = number of phases).

To be more specific, the composition of each phase is determined by $C - 1$ intensive variables (such as mole fractions) in each phase. The total number of variables is $(C - 1)P + 2$, where the extra two are temperature T and pressure p. The number of constraints is $C(P - 1)$, since the chemical potential of each component must be equal in all phases. Subtract the number of constraints from the number of variables to obtain the number of degrees of freedom as $F = (C - 1)P + 2 - C(P - 1) = C - P + 2$.

The rule is valid provided the equilibrium between phases is not influenced by gravitational, electrical or magnetic forces, or by surface area, and only by temperature, pressure, and concentration.

Consequences and Examples

Pure Substances (One Component)

For pure substances $C = 1$ so that $F = 3 - P$. In a single phase ($P = 1$) condition of a pure component system, two variables ($F - 2$), such as temperature and pressure, can be chosen independently to be any pair of values consistent with the phase. However, if the temperature and pressure combination ranges to a point where the pure component undergoes a separation into two phases ($P = 2$), F decreases from 2 to 1. When the system enters the two-phase region, it becomes no longer possible to independently control temperature and pressure.

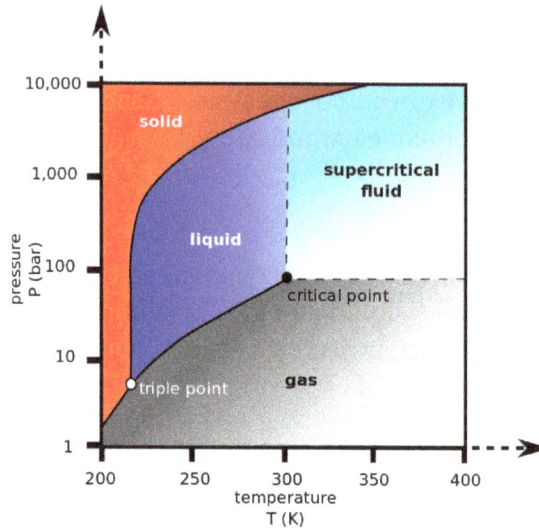

Carbon dioxide pressure-temperature phase diagram showing
the triple point and critical point of carbon dioxide

In the phase diagram to the above, the boundary curve between the liquid and gas regions maps the constraint between temperature and pressure when the single-component system has separated into liquid and gas phases at equilibrium. If the pressure is increased by compression, some of the gas condenses and the temperature goes up. If the temperature is decreased by cooling, some of the gas condenses, decreasing the pressure. Throughout both processes, the temperature and pressure stay in the relationship shown by this boundary curve unless one phase is entirely consumed by evaporation or condensation, or unless the critical point is reached. As long as there are two phases, there is only one degree of freedom, which corresponds to the position along the phase boundary curve.

The critical point is the black dot at the end of the liquid–gas boundary. As this point is approached, the liquid and gas phases become progressively more similar until, at the critical point, there is no longer a separation into two phases. Above the critical point and away from the phase boundary curve, $F = 2$ and the temperature and pressure can be controlled independently. Hence there is only one phase, and it has the physical properties of a dense gas, but is also referred to as a supercritical fluid.

Of the other two-boundary curves, one is the solid–liquid boundary or melting point curve which indicates the conditions for equilibrium between these two phases, and the other at lower temperature and pressure is the solid–gas boundary.

Even for a pure substance, it is possible that three phases, such as solid, liquid and vapour, can exist together in equilibrium ($P = 3$). If there is only one component, there are no degrees of freedom ($F = 0$) when there are three phases. Therefore, in a single-component system, this three-phase mixture can only exist at a single temperature and pressure, which is known as a triple point. Here there are two equations $\mu_{sol}(T, p) = \mu_{liq}(T, p) = \mu_{vap}(T, p)$, which are sufficient to determine the two variables T and p. In the diagram for CO_2 the triple point is the point at which the solid, liquid and gas phases come together, at 5.2 bar and 217 K. It is also possible for other sets of phases to form a triple point, for example in the water system there is a triple point where ice I, ice III and liquid can coexist.

If four phases of a pure substance were in equilibrium ($P = 4$), the phase rule would give $F = -1$, which is meaningless, since there cannot be -1 independent variables. This explains the fact that four phases of a pure substance (such as ice I, ice III, liquid water and water vapour) are not found in equilibrium at any temperature and pressure. In terms of chemical potentials there are now three equations, which cannot in general be satisfied by any values of the two variables T and p, although in principle they might be solved in a special case where one equation is mathematically dependent on the other two. In practice, however, the coexistence of more phases than allowed by the phase rule normally means that the phases are not all in true equilibrium.

Two-component Systems

For binary mixtures of two chemically independent components, $C = 2$ so that $F = 4 - P$. In addition to temperature and pressure, the other degree of freedom is the composition of each phase, often expressed as mole fraction or mass fraction of one component.

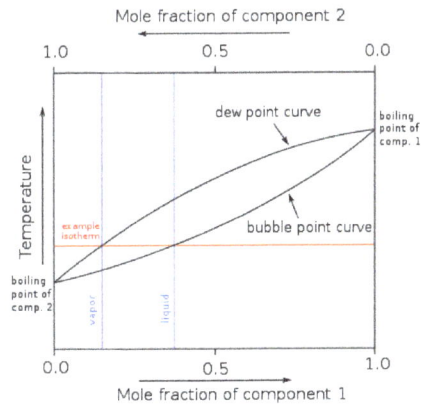

Boiling Point Diagram

As an example, consider the system of two completely miscible liquids such as toluene and benzene, in equilibrium with their vapours. This system may be described by a boiling-point diagram which shows the composition (mole fraction) of the two phases in equilibrium as functions of temperature (at a fixed pressure).

Four thermodynamic variables which may describe the system include temperature (T), pressure (p), mole fraction of component 1 (toluene) in the liquid phase (x_{1L}), and mole fraction of component 1 in the vapour phase (x_{1V}). However, since two phases are in equilibrium, only two of these variables can be independent ($F = 2$). This is because the four variables are constrained by two relations: the equality of the chemical potentials of liquid toluene and toluene vapour, and the corresponding equality for benzene.

For given T and p, there will be two phases at equilibrium when the overall composition of the system (system point) lies in between the two curves. A horizontal line (isotherm or tie line) can be drawn through any such system point, and intersects the curve for each phase at its equilibrium composition. The quantity of each phase is given by the lever rule (expressed in the variable corresponding to the x-axis, here mole fraction).

For the analysis of fractional distillation, the two independent variables are instead considered to be liquid-phase composition (x_{1L}) and pressure. In that case the phase rule implies that the equilibrium temperature (boiling point) and vapour-phase composition are determined.

Liquid–vapour phase diagrams for other systems may have azeotropes (maxima or minima) in the composition curves, but the application of the phase rule is unchanged. The only difference is that the compositions of the two phases are equal exactly at the azeotropic composition.

Phase Rule at Constant Pressure

For applications in materials science dealing with phase changes between different solid structures, pressure is often imagined to be constant (for example at one atmosphere), and is ignored as a degree of freedom, so the rule becomes

$$F = C - P + 1.$$

This is sometimes misleadingly called the "condensed phase rule", but it is not applicable to condensed systems which are subject to high pressures (for example, in geology), since the effects of these pressures can be important.

The best way to monitor the process of solidification is to measure its temperature by a thermocouple. Figure below gives a plot of temperature (T) as a function of time (t). It shows that temperature keeps dropping continuously till it reaches a temperature when cooling stops till the liquid gets transformed totally into solid. Thereafter temperature keeps dropping again.

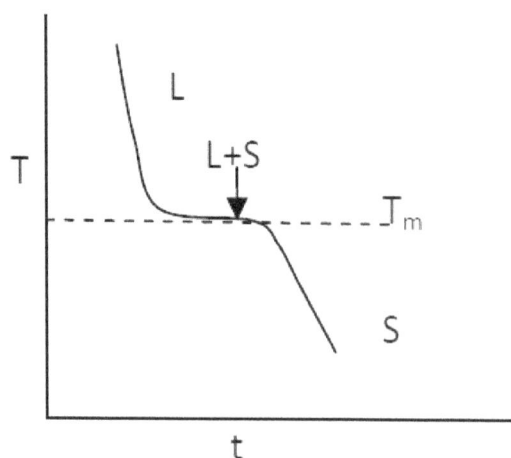

A plot of temperature (T) as a function of time (t).

Gibb's phase rule gives a simple relationship between numbers of phases (P), number of component (C), and degree of freedom (F) for a given system. This is stated as P+F = C+2. Note that C = 1 since it is a pure metal & P = 1 since there is only one phase above Tm (melting point). Therefore F = 2. It means this state can have 2 controllable variables. These are temperature and pressure.

Gibb's Free Energy (G)

Gibb's free energy (G) is a measure of the stability of a phase at a given temperature and pressure.

The lower the free energy higher is its stability. At a given temperature (T) this is defined in terms of molar enthalpy (H) and molar entropy (S) of a particular phase as follows:

$$G = H - TS$$

This can be written in differential form in terms of pressure (P) & temperature (T), the two common variables for most transformation processes, as:

$$dG = V\,dP - S\,dT$$

The equation above is valid both for solid and liquid. Let us write these explicitly for the two using subscript S (for solid) & L (for liquid).

$$dG_L = V_L\,dP - S_L\,dT \qquad \text{for liquid}$$
$$dG_S = V_S\,dP - S_S\,dT \qquad \text{for liquid}$$

At the melting point the free energy of solid and liquid should be the same. Equating equation 3 & 4 and subsequent rearrangement of the terms one gets the following expression:

$$\frac{dP}{dT} = \frac{S_L - S_S}{V_L - V_S} = \frac{T_m}{T_m}\left(\frac{\Delta S_{L-S}}{\Delta S_{L-S}}\right) = \frac{\Delta H_{L-S}}{T_m \Delta V_{L-S}}$$

Apply this to ice making. When water becomes ice there is an increase in volume and it is accompanied by release of heat.

Phase Diagram of Pure Iron

Iron can exist in more than one crystalline form. Melting point of pure iron is 1539°C. If it is cooled from its molten state it first solidifies in the form a bcc phase. The cooling curve is therefore is expected to show steps or discontinuities at these three temperatures. A step indicates coexistence of two phases. The phase diagram of pure iron is shown in figure. This shows the effect of pressure on the temperatures at which transformation of iron from one crystalline from to another takes place.

The dashed vertical line in figure below has been drawn at 1atmospheric pressure. It intersects the boundaries between alpha and gamma at 910°C. Alpha (BCC) gamma (FCC) transformation is accompanied by contraction. Therefore the slope of the line is negative. At 1atmosphere pressure gamma delta transformation takes place at 1400C whereas delta liquid transformation takes place at 1539C. Both of these are accompanied by expansion therefore the slopes of the two lines are positive.

The phase diagram of pure iron. This gives the stability of $\alpha, \gamma, \delta, L$ & G phases in various temperature pressure domains.

Real Cooling Curve: Super (Under) Cooling

In reality the transformation from liquid to solid state begins only after it has cooled below its melting point. Figure below shows a sketch of real cooling curve. Once the process initiates the latent heat that is released by the metal raises the temperature back to its melting point. Thereafter the temperature remains constant till the solidification is complete.

A real cooling curve.

Along with this G versus T plot of the solid and liquid phases have been shown. At a temperature lower than its melting point $G_S < G_L$ signifying S is more stable than L.

A spherical nucleus of solid of radius r formed in a pool of liquid.

When a solid forms in a pool of liquid a new surface is created. This has a finite energy. It is seen from above figure that $G_S < G_L$ when solidification begins. This difference in free energy acts as the driving force for solidification. Once this is large enough for a stable nucleus of solid to form the process of solidification begins. Until then unstable nuclei may appear and disappear again and again. Soon after a stable nucleus forms it keeps growing. Thus the process of solidification can be visualized as one of nucleation and growth.

This has a volume $\frac{4\pi}{3}r^3$ and surface area $4\pi r^2$. The creation of a new surface needs energy. Note that $G_S < G_L$, $DG_{L\text{-}S} < 0$. This can help do so. Assuming that surface energy / unit area is σ and free energy change /unit volume is Δf_v it is possible to derive an expression for the total change in energy for solidification.

From the details given in the above paragraph the net energy of transformation is given by the following expression.

$$\Delta f_T = \frac{4\pi}{3}r^3 \Delta f_v + 4\pi r^2 \sigma = ar^3 + br^2$$

Note that the volume free energy (or the driving force) term increases as the cube of the radius and the surface energy term increases as the square of the radius. The former is negative whereas the latter is positive. At extremely smaller values of r the surface energy term is more dominant and the net energy is positive. However as r grows the energy is likely to increase and pass through a peak. The location of the peak can be estimated by equating $\frac{d\Delta f_T}{dr} = 0$.

$$\frac{d\Delta f_T}{dr} = 4\pi r^2 \Delta f_v + 8\pi r^2 \sigma = 0; \quad \therefore r_C = -\frac{2\sigma}{\Delta f_v}$$

This gives the critical size of the nucleus. On substitution of the size of the critical nucleus in the expression for the total energy one obtains the maximum value it can have. This is given by:

$$\Delta f_{T\max} = \frac{16\pi}{3}\frac{\sigma^3}{(\Delta f_v)^2}$$

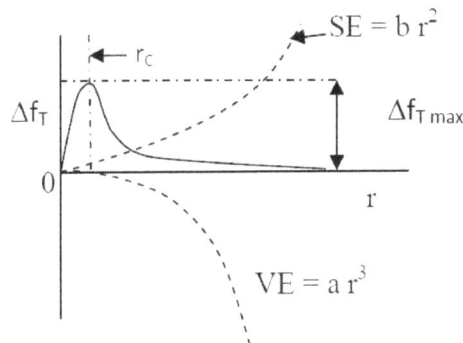

Initially SE contribution far exceeds VE so that the net energy is positive. However VE increases faster with increasing size of the nucleus. The total energy passes through a peak. The corresponding value of r is the critical nucleus size. Nuclei with r < rC are called unstable nuclei (embryo) and those with r>rC are the stable nuclei (embryo) that can grow spontaneously.

From the above analysis of the energetic of solidification the following conclusions can be drawn.

(i) Super-cooling is necessary for stable nuclei of solid to form in a pool of liquid.

(ii) Only nuclei having size greater than a critical value rc can grow spontaneously.

(iii)Higher the super cooling lower is the size of critical nucleus.

(iv) Higher the super-cooling lower is the nucleation barrier. This is also called the activation hill.

Concept of Nucleation and Growth

Solidification takes place by nucleation and growth. Atoms in a solid or in a liquid are never stationary. Assume that in a pool of molten metal there is a virtual boundary that separates a potential stable nucleus (embryo) from the surrounding liquid. Such virtual nuclei infinitesimally small in size may be assumed to be present in the liquid that is about to solidify. Atoms keep trying to cross the barrier.

$$v = v_0 \exp\left(-\frac{E}{kT}\right)$$

This shows that with increasing E the probability of formation of stable nuclei should decrease. Once stable nuclei form they would continue to grow. Initially the growth occurs at same rate in all direction until the growth is hindered due to impingement. The formation of new nuclei within the remaining liquid also continues. When the process is complete the solid is found to consist of several grains (crystals). The size of the grains may differ depending on whether it developed from a nucleus formed right in the beginning or towards the end of the process.

A schematic representation of nucleation and growth of solid nuclei during solidification from the molten liquid state.

In the above figure colours denote different orientations of grains:-

(a) Initially there are fewer nuclei. Some of them have grown.

(b) Shows that growth ceases along certain directions due to impingement. A few more nuclei have formed. All of these continue to grow.

(c) Shows a state when most of the space is filled up indicating that the process is nearly complete. Grains appear to be randomly oriented.

Grain Orientation

The process of nucleation described above is totally random. Grains may nucleate anywhere in the melt. It is also known as homogeneous nucleation. If the orientations of the grains are represented in a standard projection by their cube poles these will be located uniformly all over the entire standard projection.

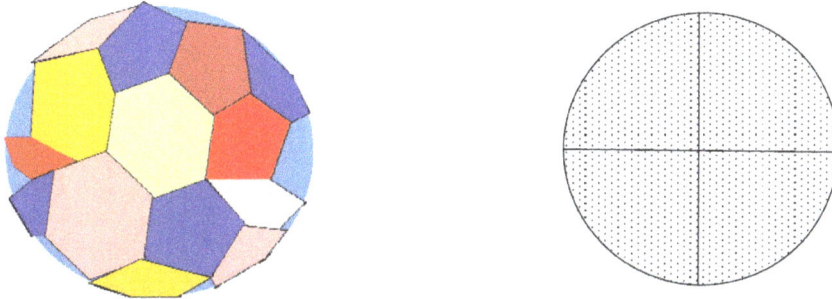

(a) Shows a typical microstructure of pure metal each grain having different orientations. Sometimes grain lustre depends on its orientation. (b) This shows distribution of <001> poles of all the grains on a standard projection. This type of diagram is known as pole figure. Homogeneous versus heterogeneous nucleation

We have so far looked at homogeneous nucleation. This is most likely to happen if the molten pool of metal has no free surface. This however is only a hypothetical case which may occur only at the centre of the mould containing the molten metal. Mould surface provides an interface where solid may nucleate. When it takes place on a pre-existing solid surface the process is known as heterogeneous nucleation. This is illustrated in fig 11. Since the mould surface provides a part of the surface energy it would need less driving force (super cooling). Thus by providing such nucleation sites it is possible to promote solidification. More the number of sites for nucleation finer are the grains. This is one of the ways of having fine grain structure on solidification. The process is known as inoculation where very small amount of easily dispersible fine solid particles are added to the metal as it is poured into mould.

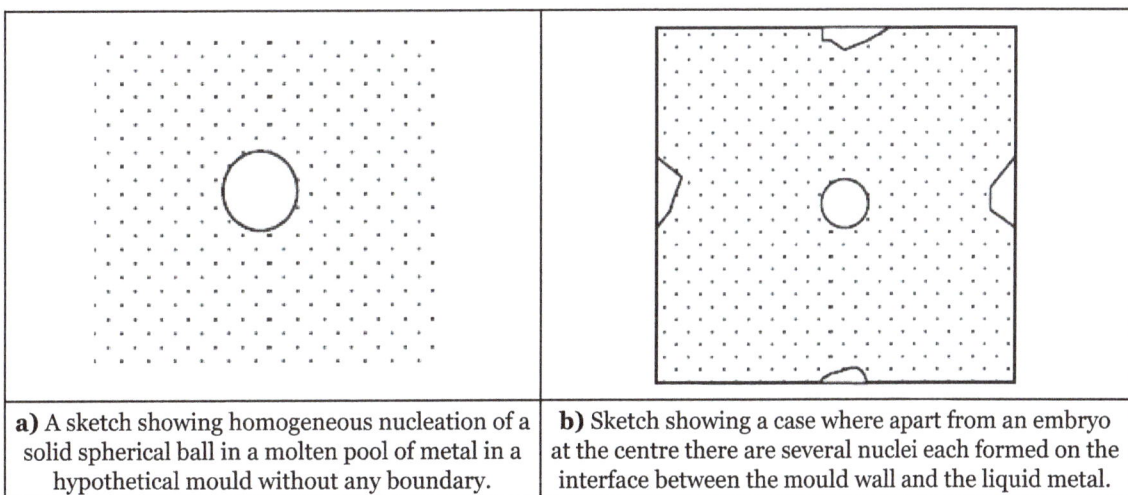

a) A sketch showing homogeneous nucleation of a solid spherical ball in a molten pool of metal in a hypothetical mould without any boundary.	b) Sketch showing a case where apart from an embryo at the centre there are several nuclei each formed on the interface between the mould wall and the liquid metal.

Figure above illustrates the role played by the respective surface tension (or surface energy; both represent the same physical parameter) and the contact angle. In this case the mould surface acts as a substrate on which an embryo of the solid nucleates. At the point of contact there are three surface forces; one between the mould and the embryo (σ_{ME}), the second between the mould & the liquid (σ_{ME}) and the third between the embryo & the liquid (σ_{EL}). The angle between σ_{EL} and σ_{ME} is known as the contact angle. Invoking the condition of equilibrium between the three forces it is possible to find out the condition that would promote nucleation.

$$\sigma_{ML} = \sigma_{ME} + \sigma_{EL}\cos\theta \quad \therefore \cos\theta = \frac{\sigma_{ML} - \sigma_{ME}}{\sigma_{EL}}$$

A sketch illustrating the role played by the surface tensions between the three surfaces that come in contact during the nucleation of an embryo.

Clearly the condition (a) in fig above is the favourable condition for nucleation. In the case (b) where the contact angle > 90 is not favourable for nucleation. The selection of inoculating agent to get a fine grain structure after solidification is based on this simple concept.

Directional Solidification

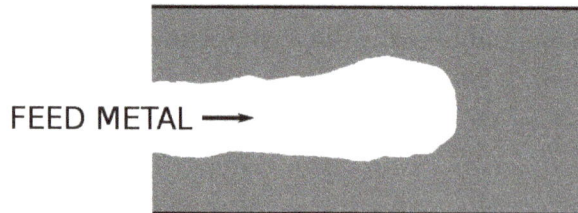

Directional solidification

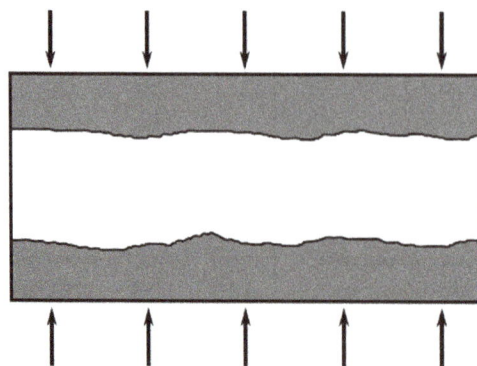

Progressive solidification

Directional solidification (DS) and progressive solidification are types of solidification within castings. Directional solidification is solidification that occurs from farthest end of the casting and works its way towards the sprue. Progressive solidification, also known as parallel solidification, is solidification that starts at the walls of the casting and progresses perpendicularly from that surface.

Theory

Most metals and alloys shrink as the material changes from a liquid state to a solid state. Therefore, if liquid material is not available to compensate for this shrinkage a *shrinkage defect* forms. When progressive solidification dominates over directional solidification a shrinkage defect will form.

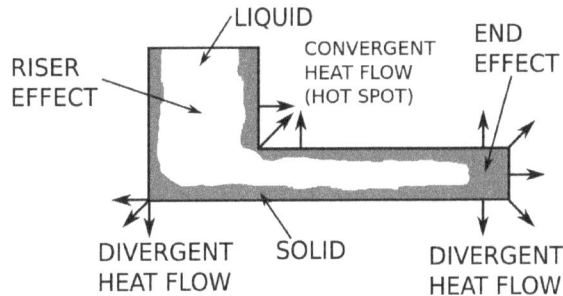

The geometrical shape of the mold cavity has direct effect on progressive and directional solidification. At the end of tunnel-type geometries divergent heat flow occurs, which causes that area of the casting to cool faster than surrounding areas; this is called an *end effect*. Large cavities do not cool as quickly as surrounding areas because there is less heat flow; this is called a *riser effect*. Also note that corners can create divergent or convergent (also known as *hot spots*) heat flow areas.

In order to induce directional solidification chills, risers, insulating sleeves, control of pouring rate, and pouring temperature can be utilized.

Directional solidification can be used as a purification process. Since most impurities will be more soluble in the liquid than in the solid phase during solidification, impurities will be "pushed" by the solidification front, causing much of the finished casting to have a lower concentration of impurities than the feedstock material, while the last solidified metal will be enriched with impurities. This last part of the metal can be scrapped or recycled. The suitability of directional solidification in removing a specific impurity from a certain metal depends on the partition coefficient of the impurity in the metal in question, as described by the Scheil equation. Directional solidification is frequently employed as a purification step in the production of multicrystalline silicon for solar cells.

How the heat is extracted from the molten pool of metal is of considerable importance in determining the structure of the solid. When liquid metal is poured into a mould (sometime preheated) the heat flows from the liquid through the mould wall to the surrounding environment. Heat flows from higher to lower temperature. The rate is proportional to thermal gradient. It also depends on the effective heat transfer coefficient of the entire system consisting of the liquid, solidified layer, the mould and the region surrounding the mould. Figure illustrates a case where heat is being extracted in one direction from the liquid metal through solid metal and the mould. Note that the solid liquid interface is flat in this case. The liquid to solid transformation is accompanied by evolution of heat at the interface. This can only be extracted down the temperature gradient.

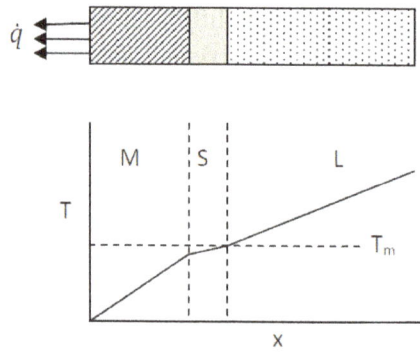

Figure showing the direction of heat flow from the liquid metal through the solid and the mould.
It also gives the temperature as a function of distance. The S/L interface is likely to be flat.
The rate at which it moves would depend on how fast the heat can be extracted.

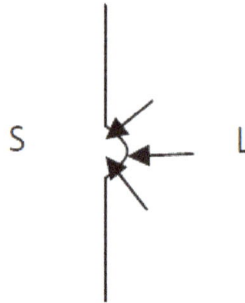

Figure showing a bulge in the S/L interface. The arrow denotes the direction of heat flow. This makes
it unstable as the heat flowing into the bulge would melt it again.

Deformation (Engineering)

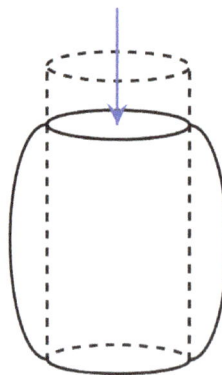

Compressive stress results in deformation which shortens the
object but also expands it outwards.

In materials science, deformation refers to any changes in the shape or size of an object due to-

- an applied force (the deformation energy in this case is transferred through work) or
- a change in temperature (the deformation energy in this case is transferred through heat).

The first case can be a result of tensile (pulling) forces, compressive (pushing) forces, shear, bending or torsion (twisting).

In the second case, the most significant factor, which is determined by the temperature, is the mobility of the structural defects such as grain boundaries, point vacancies, line and screw dislocations, stacking faults and twins in both crystalline and non-crystalline solids. The movement or displacement of such mobile defects is thermally activated, and thus limited by the rate of atomic diffusion.

Deformation is often described as strain.

As deformation occurs, internal inter-molecular forces arise that oppose the applied force. If the applied force is not too great these forces may be sufficient to completely resist the applied force and allow the object to assume a new equilibrium state and to return to its original state when the load is removed. A larger applied force may lead to a permanent deformation of the object or even to its structural failure.

In the figure it can be seen that the compressive loading (indicated by the arrow) has caused deformation in the cylinder so that the original shape (dashed lines) has changed (deformed) into one with bulging sides. The sides bulge because the material, although strong enough to not crack or otherwise fail, is not strong enough to support the load without change, thus the material is forced out laterally. Internal forces (in this case at right angles to the deformation) resist the applied load.

The concept of a rigid body can be applied if the deformation is negligible.

Types of Deformation

Depending on the type of material, size and geometry of the object, and the forces applied, various types of deformation may result. The image to the right shows the engineering stress vs. strain diagram for a typical ductile material such as steel. Different deformation modes may occur under different conditions, as can be depicted using a deformation mechanism map.

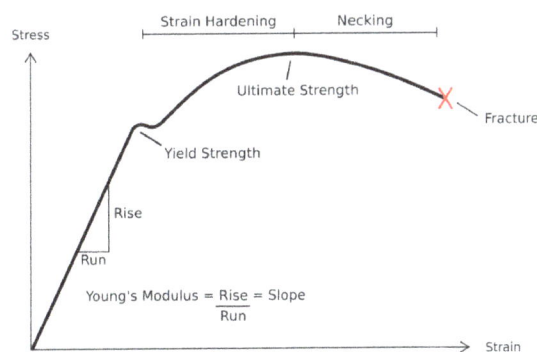

Typical stress vs. strain diagram indicating the various stages of deformation.

Elastic Deformation

This type of deformation is reversible. Once the forces are no longer applied, the object returns to its original shape. Elastomers and shape memory metals such as Nitinol exhibit large elastic deformation ranges, as does rubber. However elasticity is nonlinear in these materials. Normal metals, ceramics and most crystals show linear elasticity and a smaller elastic range.

Linear elastic deformation is governed by Hooke's law, which states:

$$\sigma = E\varepsilon$$

Where σ is the applied stress, E is a material constant called Young's modulus or elastic modulus, and ε is the resulting strain. This relationship only applies in the elastic range and indicates that the slope of the stress vs. strain curve can be used to find Young's modulus (E). Engineers often use this calculation in tensile tests. The elastic range ends when the material reaches its yield strength. At this point plastic deformation begins.

Note that not all elastic materials undergo linear elastic deformation; some, such as concrete, gray cast iron, and many polymers, respond in a nonlinear fashion. For these materials Hooke's law is inapplicable.

Plastic Deformation

This type of deformation is irreversible. However, an object in the plastic deformation range will first have undergone elastic deformation, which is reversible, so the object will return part way to its original shape. Soft thermoplastics have a rather large plastic deformation range as do ductile metals such as copper, silver, and gold. Steel does, too, but not cast iron. Hard thermosetting plastics, rubber, crystals, and ceramics have minimal plastic deformation ranges. An example of a material with a large plastic deformation range is wet chewing gum, which can be stretched dozens of times its original length.

Under tensile stress, plastic deformation is characterized by a strain hardening region and a necking region and finally, fracture (also called rupture). During strain hardening the material becomes stronger through the movement of atomic dislocations. The necking phase is indicated by a reduction in cross-sectional area of the specimen. Necking begins after the ultimate strength is reached. During necking, the material can no longer withstand the maximum stress and the strain in the specimen rapidly increases. Plastic deformation ends with the fracture of the material.

Metal Fatigue

Another deformation mechanism is metal fatigue, which occurs primarily in ductile metals. It was originally thought that a material deformed only within the elastic range returned completely to its original state once the forces were removed. However, faults are introduced at the molecular level with each deformation. After many deformations, cracks will begin to appear, followed soon after by a fracture, with no apparent plastic deformation in between. Depending on the material, shape, and how close to the elastic limit it is deformed, failure may require thousands, millions, billions, or trillions of deformations.

Metal fatigue has been a major cause of aircraft failure, especially before the process was well understood. There are two ways to determine when a part is in danger of metal fatigue; either predict when failure will occur due to the material/force/shape/iteration combination, and replace the vulnerable materials before this occurs, or perform inspections to detect the microscopic cracks and perform replacement once they occur. Selection of materials not likely to suffer from metal fatigue during the life of the product is the best solution, but not always possible. Avoiding shapes with sharp corners limits metal fatigue by reducing stress concentrations, but does not eliminate it.

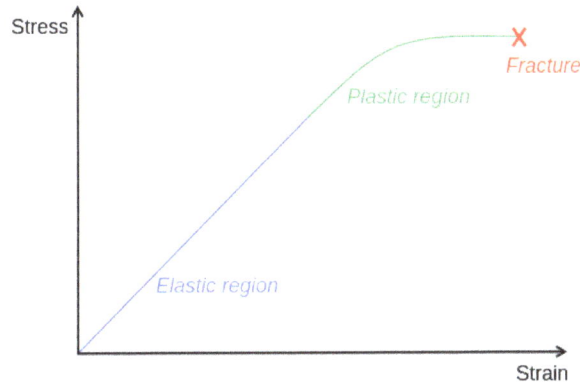

Diagram of a stress–strain curve, showing the relationship between stress (force applied) and strain (deformation) of a ductile metal.

Compressive Failure

Usually, compressive stress applied to bars, columns, etc. leads to shortening.

Loading a structural element or specimen will increase the compressive stress until it reaches its compressive strength. According to the properties of the material, failure modes are yielding for materials with ductile behavior (most metals, some soils and plastics) or rupturing for brittle behavior (geomaterials, cast iron, glass, etc.).

In long, slender structural elements — such as columns or truss bars — an increase of compressive force F leads to structural failure due to buckling at lower stress than the compressive strength.

Fracture

This type of deformation is also irreversible. A break occurs after the material has reached the end of the elastic, and then plastic, deformation ranges. At this point forces accumulate until they are sufficient to cause a fracture. All materials will eventually fracture, if sufficient forces are applied.

Misconceptions

A popular misconception is that all materials that bend are "weak" and those that don't are "strong". In reality, many materials that undergo large elastic and plastic deformations, such as steel, are able to absorb stresses that would cause brittle materials, such as glass, with minimal plastic deformation ranges, to break.

When a material is subjected to stress depending on whether it is tensile or compressive the distance between atoms either increase or decrease. If it comes back to its initial state when the stress is withdrawn it is known as elastic deformation. It does not leave any sign of change in shape and size. As against this when the stress exceeds the yield strength of the material it leaves behind permanent signs of deformation. In this section we shall learn about the representation of stresses & strains and their relationship. The number of elastic constants needed to describe these depends on the crystal structure. This is connected with the crystal symmetry. The concept of isotropy & anisotropy will be introduced. We shall learn also about the basic difference between elastic & plastic deformation.

Elastic Deformation

Metals are crystalline. A normal solid is made of several grains. When it is subjected to tensile or compressive stress each of the grains undergoes identical deformation. Grains are made of regular arrays of atoms. However the orientation of crystal planes and directions may vary from grain to grain. In order to understand what happens during elastic deformation let us look at the atomic array in one isolated crystal. This is illustrated in fig below in the form of a set of sketches showing the effect of uniaxial stress.

Physical Character & Representation of Stresses & Strains at a Point

Stress is defined as load (or force) per unit area. Note that both force and area are vectors. Both have magnitude and direction. Let us represent the force F in terms of three normal components F_1, F_2, & F_3 along the Cartesian axes X1, X2, X3. In general a component of force can be represented as Fi where the subscript i can have values 1, 2 or 3 depending on whether it acts along axis X_1, X_2, X_3. Likewise the components of an area A can be represented as Ai. This means that a component of a vector is represented by one subscript. Mathematically the stress relates two vectors (force and area). The best way to represent this is as follows:

$$\overline{F} = \overline{\overline{\sigma}}\,\overline{A} \text{ Or } \{F\} = [\sigma]\{A\} \text{ or } F_i = \sigma_{ij}A_j \qquad (1)$$

Note that to represent stress two subscripts have been used. Force & area vectors may be considered as 1x3 matrices. To connect these two one needs a 3x3 matrix. This is the correct description of stress. It is a second order tensor. All the three forms of equation 1 are same. The last expression uses repeated subscript j on its right hand side. This denotes summation over the subscript j. This type of representation is not only short but also helps conversion of stress state from one reference axes to another.

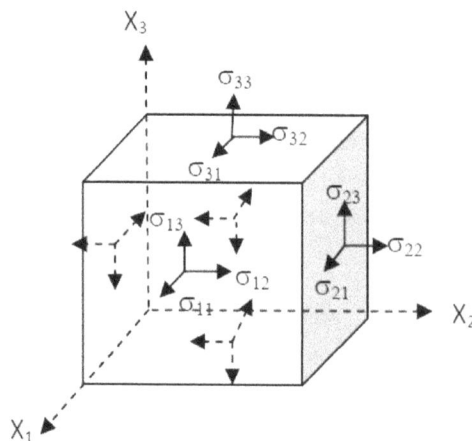

The figure illustrates the representation of stress at a point by visualizing a small imaginary cube around it. Note the use of two subscripts used to denote stress. σ_{11} is stress acting on plane perpendicular to X_1 and along direction X_1. σ_{12} is stress acting on the same plane but along X_2. The former is a normal stress whereas the latter is a shear stress. Unequal subscripts denote shear stress. The stress at a point consists of 3 normal and a pair of 3 shear stresses.

The state of stress around a point is thus represented mathematically as follows:

$$\begin{bmatrix} \sigma_{11} & \sigma_{12} & \sigma_{13} \\ \sigma_{21} & \sigma_{22} & \sigma_{23} \\ \sigma_{31} & \sigma_{32} & \sigma_{33} \end{bmatrix} \qquad (2)$$

Like stress strain too at a point can be represented by a second order symmetric tensor as given below:

$$\begin{bmatrix} \varepsilon_{11} & \varepsilon_{12} & \varepsilon_{13} \\ \varepsilon_{21} & \varepsilon_{22} & \varepsilon_{23} \\ \varepsilon_{31} & \varepsilon_{32} & \varepsilon_{33} \end{bmatrix} \qquad (3)$$

Note if the two subscripts of stress (or strain) are identical then it denotes normal stress (it in a direction perpendicular to the plane).

		Old axes		
		OX_1	OX_2	OX_3
New axes	OX'_1	a_{11}	a_{12}	a_{13}
	OX'_2	a_{21}	a_{22}	a_{23}
	OX'_3	a_{31}	a_{32}	a_{33}

The stress state with respect to the set of axes can be obtained using the following rules of transformation:

$$\sigma'_{ij} = a_{ik} a_{jl} \sigma_{kl} \qquad \text{for conversion from old to new reference axes} \quad (4)$$

$$\sigma_{kl} = a_{ik} a_{jl} \sigma'_{ij} \qquad \text{for conversion from new to old reference axes} \quad (5)$$

One 2 fold Axis of Symmetry

Let us now look at the implication of symmetry elements on the number of independent elastic constants. Take the case of one 2 fold axis of symmetry. Earlier figure shows two sets of reference axes. The primed axes have been obtained by a rotation of 180° (π) about X3. Note the direction cosine matrix for the new axes with respect to the old one.

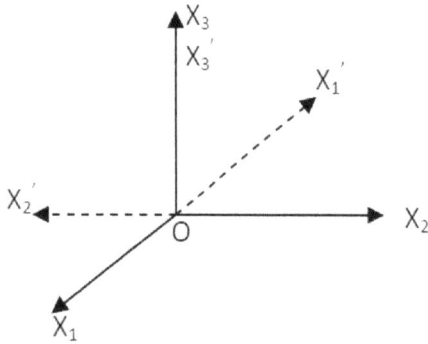

	X_1	X_2	X_3
X_1'	-1	0	0
X_2'	0	-1	0
X_3'	0	0	1

The orientation relation between the old axes (OX_1, OX_2, OX_3) and the new axes (OX_1', OX_2', OX_3')

This also includes the direction cosine matrix. Since this happens to be an axis of symmetry $C_{ij} = C_{ij}' \ \& \ S_{ij} = S_{ij}'$.

Therefore stress strain relation with respect to either reference axes can be written a $\sigma_i = C_{ij}\varepsilon_j \ \& \ \sigma_i' = C_{ij}\varepsilon_j'$. For example σ_1 in contracted notation is σ_{11}.

Thus: $\sigma_1' = \sigma_{11}' = a_{1k}a_{1l}\sigma_{kl} = (-1)(-1)\sigma_{11} = \sigma_1$. Similar relation would hold for the strain tensor as well. In general it can be described as $\sigma_k' = \sigma_k$ and $\varepsilon_k' = \varepsilon_k$ for k = 1, 2, 3 & 6. Repeat the above steps to show that $\sigma_k' = -\sigma_k$ and $\varepsilon_k' = -\varepsilon_k$ for k = 4 & 5.

$$\sigma_1 = C_{11}\varepsilon_1 + C_{12}\varepsilon_2 + C_{13}\varepsilon_3 + C_{14}\varepsilon_4 + C_{15}\varepsilon_5 + C_{16}\varepsilon_6 \qquad (6)$$
$$\sigma_1' = C_{11}\varepsilon_1 + C_{12}\varepsilon_2 + C_{13}\varepsilon_3 - C_{14}\varepsilon_4 + C_{15}\varepsilon_5 + C_{16}\varepsilon_6 \qquad (7)$$

The two should be identical because of symmetry. This is possible if $C_{14} = C_{15} = 0$.

Two 2 fold axis of symmetry (Orthorhombic crystal / orthotropic material):

Assume that in addition to X3, X1 too is a 2 fold axis of symmetry. Figure below represents the orientation relation between the two sets of axes. It also indicates the direction cosine matrix.

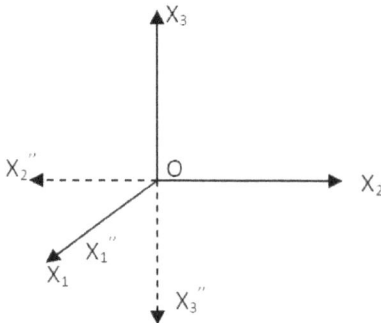

	X_1	X_2	X_3
X_1''	1	0	0
X_2''	0	-1	0
X_3''	0	0	-1

The orientation relations between two sets of axes.

Here OX1 is an axis of 2 fold symmetry. Using the direction cosine matrix it can be shown that as $\sigma_k' = \sigma_k$ and $\varepsilon_k' = \varepsilon_k$. for k = 1, 2, 3 & 4. Repeat the above steps to show that $\sigma_k' = -\sigma_k$ and $\varepsilon_k' = -\varepsilon_k$ for k =5 & 6.

Since X_3 is also an axis of 2 fold symmetry more numbers of elastic constants must be equal to 0.

Using the transformation rule introduced here it can be shown that:

$$\sigma_1 = C_{11}\varepsilon_1 + C_{12}\varepsilon_2 + C_{13}\varepsilon_3 + C_{16}\varepsilon_6 \qquad (8)$$

$$\sigma_1' = C_{11}\varepsilon_1 + C_{12}\varepsilon_2 + C_{13}\varepsilon_3 - C_{16}\varepsilon_6 \qquad (9)$$

The two equations must give the same value for σ_1. Thus $C_{16} = 0$ Following the same logic it can be shown that $C_{26} = C_{36} = 0$. Equate σ_1 and σ_4' using the following expressions:

$$\sigma_1 = C_{44}\varepsilon_4 + C_{45}\varepsilon_5 \qquad (10)$$

$$\sigma_4' = C_{44}\varepsilon_4 - C_{45}\varepsilon_5 \qquad (11)$$

The two are equal only if $C_{45} = 0$. Thus the number of elastic constants for orthorhombic crystal is reduced to 9.

Additionally if there is at least one 4 fold axis of symmetry it can easily be shown that the number of nonzero elastic constants reduces further.

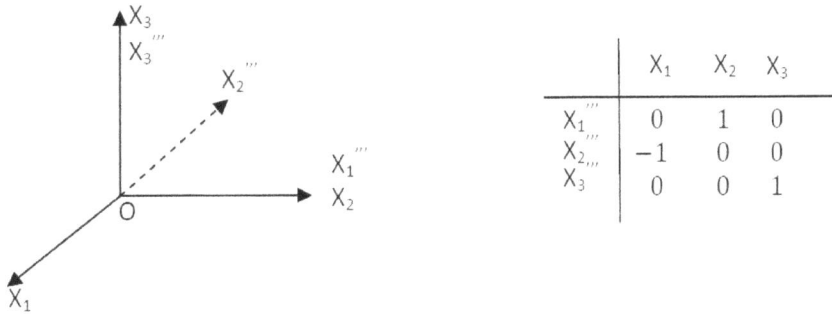

	X_1	X_2	X_3
$X_{1'''}$	0	1	0
$X_{2'''}$	−1	0	0
X_3	0	0	1

Illustrates the orientation relation between new and old axes if OX_3 is also a 4 fold axis of symmetry. The angle of rotation is $\pi/4$. The corresponding direction cosine matrix between the old (OX) and the new (OX''') axes are also given.

Using the direction cosine matrix given in the above figure it can be shown that $\sigma_1'' = \sigma_2$; $\sigma_2'' = \sigma_1$; $\sigma_3'' = \sigma_3$; $\sigma_4'' = -\sigma_5$; $\sigma_5'' = \sigma_4$; $\sigma_6'' = -\sigma_6$; and

$\varepsilon_1'' = \varepsilon_2$; $\varepsilon_2'' = \varepsilon_1$; $\varepsilon_3'' = \varepsilon_3$; $\varepsilon_4'' = -\varepsilon_5$; $\varepsilon_5'' = \varepsilon_4$; $\varepsilon_6'' = -\varepsilon_6$.   Thus:

$$\sigma_3 = C_{31}\varepsilon_1 + C_{32}\varepsilon_2 + C_{33}\varepsilon_3 = \sigma_3'' = C_{31}\varepsilon_1'' + C_{32}\varepsilon_2'' + C_{33}\varepsilon_3'' = C_{31}\varepsilon_2 + C_{32}\varepsilon_1 + C_{33}\varepsilon_3 \qquad (12)$$

This is possible if $C_{31} = C_{32}$ (also note that C matrix is symmetric).

Similarly $\sigma_1 = C_{11}\varepsilon_1 + C_{12}\varepsilon_2 + C_{13}\varepsilon_3 = \sigma_2'' = C_{21}\varepsilon_2 + C_{22}\varepsilon_1 + C_{33}\varepsilon_3 \qquad (13)$

This is possible if $C_{11} = C_{22}$. This procedure can be extended to show that cubic crystal has only 3 independent elastic constants. The relation between these can be represented as follows: $C_{11} = C_{22} = C_{33}; C_{12} = C_{13} = C_{23}; C_{44} = C_{55} = C_{66}$.

Therefore the elastic stiffness and elastic compliance matrix for cubic crystal can be written as:

$$
\begin{bmatrix}
C_{11} & C_{12} & C_{12} & 0 & 0 & 0 \\
C_{12} & C_{11} & C_{12} & 0 & 0 & 0 \\
C_{12} & C_{12} & C_{11} & 0 & 0 & 0 \\
0 & 0 & 0 & C_{44} & 0 & 0 \\
0 & 0 & 0 & 0 & C_{44} & 0 \\
0 & 0 & 0 & 0 & 0 & C_{44}
\end{bmatrix}
\qquad
\begin{bmatrix}
S_{11} & S_{12} & S_{12} & 0 & 0 & 0 \\
S_{12} & S_{11} & S_{12} & 0 & 0 & 0 \\
S_{12} & S_{12} & S_{11} & 0 & 0 & 0 \\
0 & 0 & 0 & S_{44} & 0 & 0 \\
0 & 0 & 0 & 0 & S_{44} & 0 \\
0 & 0 & 0 & 0 & 0 & S_{44}
\end{bmatrix}
$$

Relation between C & S

$$[C] = [S]^{-1}$$

The product of the two is a unit matrix.

Stiffness matrix   Compliance matrix

Relation between C & S $[C]=[S]^{-1}$. The product of the two is a unit matrix.

Note that there is an inverse relation between the two matrices. The product of the two is a unit matrix. This can be used to establish the following relations for elastic constants of cubic crystal.

$$S_{11} = \frac{C_{11} + C_{12}}{(C_{11} - C_{12})(C_{11} + 2C_{12})} \tag{14}$$

$$S_{44} = \frac{1}{C_{44}} \tag{15}$$

$$S_{12} = \frac{-C_{12}}{(C_{11} - C_{12})(C_{11} + 2C_{12})} \tag{16}$$

Derivation

The first two terms of the first row of the product of [C][S] for a cubic crystal are as follows:

$$C_{11}S_{11} + 2C_{12}S_{12} = 1 \tag{17}$$
$$C_{12}S_{11} + (C_{11} + C_{12})S_{12} = 0 \tag{18}$$

Solve the two simultaneous equations 17 & 18 to get equation 15 & 16. Multiply the fourth row of [C] with the fourth column of [S] to get

$$C_{44}S_{44} = 1 \therefore S_{44} = \frac{1}{C_{44}} \tag{19}$$

Elastic deformation is temporary. It disappears when the stress is withdrawn, as against this plastic deformation is permanent. It occurs only when the stress exceeds a characteristic value called yield strength. This is a material property. In this section we shall talk about the characteristics of plastic deformation; its mechanisms and its effects on the structure & properties of metal. Pure metal is ductile and malleable. It can be rolled into thin foils or drawn into fine wires. Aluminum foils for packaging and fine copper wires for electrical conductors could be made primarily because of this unique feature of metals. Metals are made of several grains (crystals) that are oriented at random. In spite of the difference in the orientation between two neighboring grains there exists a strong bond between these. When it undergoes plastic deformation the shape of the grain and the boundary change but the continuity across the grain is maintained. There is little change in its volume or crystal structure. This is one of the major differences between elastic and plastic deformation.

Deformation of Single Crystal

There are two ways a crystal can deform without any change in its crystal structure. These are slip and twin. If the stress is less than the yield strength of the metal the atoms are just pulled apart. The lattice parameter increases along the direction of stress. When it exceeds the yield strength a part of the crystal slips over the other. This takes place on a plane on which the shear stress is the maximum.

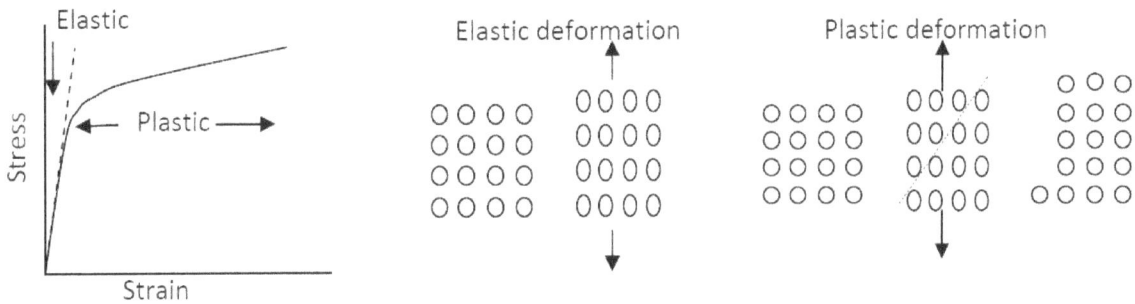

Illustrates the difference between elastic & plastic deformation.

Atoms move apart along the direction of stress but come closer along the direction perpendicular to the applied stress during elastic deformation. When the stress is withdrawn the atoms come back to their previous positions. Whereas during plastic deformation the atoms on either sides of a plane on which the shear stress is the maximum slide over one another. Displacement occurs in multiples of atomic distance on the slip plane along the slip direction. The atoms do not come back to their initial positions when the stress is withdrawn. The stress strain plot starts deviating from linearity when plastic deformation sets in.

Plastic deformation by slip can occur only on certain planes and along specific directions on the plane. The combinations of slip planes and directions on which slip can take place are known as slip system. It depends on the crystal structure. Usually the close packed planes are the slip planes since they happen to be the most widely spaced planes. Slip directions are the close packed directions. Table below gives the indices of slip planes and directions for the three most common crystal structures of metals.

Table: Lists of slip system for 3 most common crystal structures

Crystal lattice	Slip plane	Slip Direction	No. of slip system
FCC	{111}	<110>	12
HCP	{0001}	$<2\bar{1}10>$	3
BCC	{110} {112} {123}	<111>	48

It is worth noting that slip can occur only due to stress acting on the slip plane along the slip direction. Even if the applied stress is tensile along a given direction one has to find its component along the slip direction on the slip plane to know if it can induce plastic deformation. Such stresses are known as resolved shear stress. Slip occurs when the magnitude of this resolved shear stress exceeds a critical value. This is known as the critical resolved shear stress. Slip can also be visualized as a simple shear.

Before deformation   After deformation   Polishing removes slip trace

Figure showing atomic arrangement before and after deformation by slip (simple shear).

The process would leave a slip trace (step) on the surface. Atoms on both sides of the slip plane are arranged in identical fashion. If the top surface is polished the sign of deformation is totally removed. The best way to see the signs of deformation due to slip is to take a polished sample, deform and look under microscope without tampering (or polishing the surface) the surface.

The bulk of plastic deformation takes place by slip. The main features of deformation by this mechanism are:-

(i)   no change in crystal structures

(ii) no change in volume or lattice parameter

(iii) no change in crystal orientation

(iii) all the atoms above the slip plane move by identical distance on the slip plane along the slip direction

(iv) the net displacement is in multiples of atomic spacing.

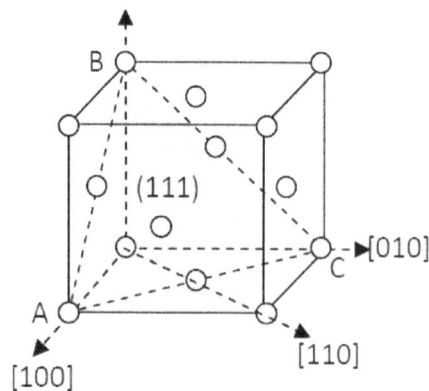

Sketch showing the location of atoms in an FCC unit cell.

Look at the close packed plane (111) and the indices of the lines joining atoms at face centers with those at the corners. All of these are close packed directions. There are 3 close packed directions in one slip plane. Since there are 4 close packed planes in FCC crystal the total number of slip planes = 4 x 3 =12.

Shapes of the grains and the sample. There is no change in volume. Deformation leaves marks on the grains as shown. These are the traces of the slip planes (lines of intersection of the slip plane and the top face of the sample). If the top face is polished the traces would vanish.

Numbers of Independent Slip Systems

Changes in shape or deformation can only take place by shear or slip (both are same). A polycrystalline metal is known to have excellent ability to deform into any desired shape. This is possible if all the crystals of which this is made can undergo any arbitrary deformation. How many slip systems a crystal must possess to satisfy this? To answer this we need to know about various possible components of strains. The total strain (ε_{ij}) is made of elastic (ε_{ij}^e) and plastic (ε_{ij}^p) components.

$$\varepsilon_{ij} = \varepsilon_{ij}^e + \varepsilon_{ij}^p$$

Let us consider the plastic strain only since we are looking at large deformation. The magnitude of elastic strain is always negligibly small in comparison to that of plastic strain. Strain is a second rank symmetric tensor. It has 6 components as shown by equation below.

$$\begin{bmatrix} \varepsilon_{11}^p & \varepsilon_{12}^p & \varepsilon_{13}^p \\ \varepsilon_{12}^p & \varepsilon_{22}^p & \varepsilon_{23}^p \\ \varepsilon_{13}^p & \varepsilon_{23}^p & \varepsilon_{33}^p \end{bmatrix}$$

The sum of the diagonal element represents total volume strain. This is an invariant. It means that the volume strain does not depend on the choice of reference axes used to represent the strain at a point.

However there is no change in volume during plastic deformation. This suggests that:

$$\varepsilon_{11}^e + \varepsilon_{22}^p + \varepsilon_{33}^p = 0$$

The number of independent strain components is therefore 5. Each of these can occur due to slip on a slip specific system. This is why 5 independent slip systems are necessary for any arbitrary deformation. Since both FCC & BCC metals have several slip systems to choose from they have excellent ductility. Gold, silver & aluminum can be rolled down to extremely thin foils. All of them have FCC structure.

Problem arises in the case of metals having HCP structure (Zn, Mg, Zr, Ti). Slip takes place only on the basal plane along any of the three close packed directions. It may be noted that vector addition of the two slip directions gives the third. Thus only two of the 3 slip systems can be considered to be independent. In fact ideal HCP structures have relatively poor ductility. In case of deviations from ideal HCP structure slip is known to occur on prism & pyramid planes. However the slip direction is still the same. This provides additional choice of slip systems from which selection of 5 independent slip systems becomes possible.

Apart from slip there is an altogether different mechanism of plastic deformation. This is called twinning. This too occurs on a specific crystal plane in such a fashion that the deformed part becomes a mirror image of the parent crystal. Unlike slip the movement or atomic displacement is a function of its distance from the twinning plane. Like slip this too occurs due to shear stress.

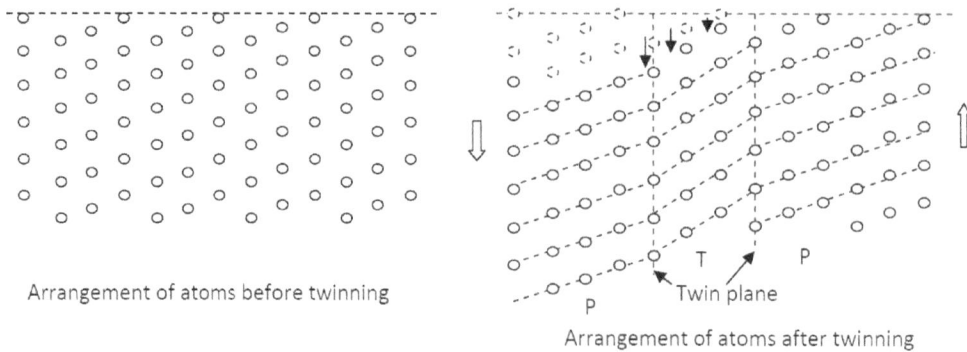

| Arrangement of atoms before twinning | Arrangement of atoms after twinning |

Figure illustrating how the atomic arrangement changes on either sides of a twin plane after deformation.

Left hand of the figure shows the positions of atoms before twinning. The right hand of the figure shows the positions of atoms after twinning. This also shows previous locations of atoms by dotted circles. These are the locations where there are no atoms now. Note the arrangement of atoms on either side of the twin planes. It exhibits reflection symmetry. Dotted lines indicate one of the close packed directions. These are differently oriented within the twin. This shows that twinning is accompanied by a change in crystal orientation. Therefore on polishing the sign of twin still remains.

Table gives a list of twin plane and direction for the 3 most common crystal structures of metals. The twin plane and direction remain undistorted. However there is a change in the orientations of the other planes and directions within the twin.

Table: List of twin planes & directions

Crystal lattice	Twin plane	Twin Direction
FCC	$\{111\}$	<110>
HCP	$\{0001\}$	<111>
BCC	$\{10\bar{1}2\}$	$<10\bar{1}1>$

Deformation during twinning takes place only within the twinned part of the crystal. However the magnitude of deformation is large. For example the magnitude of displacement in FCC crystal is $\frac{a}{6}<112>$ over a distance of $\frac{a}{3}<111>$. Therefore the magnitude of shear strain is given by:

$$\gamma = \frac{\frac{a}{6}\sqrt{1+1+4}}{\frac{a}{3}\sqrt{1+1+1}} = \sqrt{\frac{3}{6}} = 0.707$$

Unlike slip twinning is noisy and fast. One of the best examples of such a feature is that of tin cry (It has BCT structure). You can hear the noise it makes tin makes during deformation. The stress strain plot also shows serration. It is not smooth. HCP crystal has limited slip system. It needs additional deformation modes. The orientation of the deformed region is different from the matrix. As a result it may now be favorably oriented to undergo slip. Therefore twinning is very common in HCP crystal. Twinning (deformation) also takes place in metals having BCC structure. However it is not so common during plastic deformation of FCC crystal. Table below gives a comparison between the two modes of plastic deformation.

Table: A comparison of slip & twin the two modes of plastic deformation

Features	Slip	Slip
Deformation rate	Slow (time ~ milliseconds)	Fast (time ~ microseconds)
Stress – strain plot	Smooth	Shows serrations
Acoustic emission (sound)	Quiet	Noisy
Displacement of atoms	Same for all planes above the slip plane	It is proportional to its distance from the twin plane
Orientation relation	Identical on both sides of the slip plane	Twinned part is a mirror image of the parent crystal
Effect of polishing on the signs of deformation on the surface	Disappears	Does not disappear

Single crystal deformation by slip (glide): The slip takes place in a crystal when it is pulled in tension. The crystal glides on a plane the tensile axis is likely to shift. However the the grips of the testing machine will not allow it to happen. The crystal would therefore rotate. The indices of the tensile axis would change.

Critical Resolved Shear Stress

Critical resolved shear stress (CRSS) is the component of shear stress, resolved in the direction of slip, necessary to initiate slip in a grain. Resolved shear stress (RSS) is the shear component of an applied tensile or compressive stress resolved along a slip plane that is other than perpendicular or parallel to the stress axis. The RSS is related to the applied stress by a geometrical factor, m, typically the Schmid factor :

$$\tau_{RSS} = \sigma_{app}m = \sigma_{app}(\cos\phi\cos\lambda)$$

where σ_{app} is the magnitude of the applied tensile stress, Φ is the angle between the normal of the slip plane and the direction of the applied force, and λ is the angle between the slip direction and the direction of the applied force. The Schmid Factor is most applicable to FCC single crystal metals, but for polycrystal metals the Taylor factor has been shown to be more accurate. The CRSS is the value of resolved shear stress at which yielding of the grain occurs, marking the onset of plastic deformation. CRSS, therefore, is a material property and is not dependent on the applied load or grain orientation. The CRSS is related to the observed yield strength of the material by the maximum value of the Schmid factor:

$$\sigma_y = \frac{\tau_{CRSS}}{m_{max}}$$

CRSS is a constant for crystal families. Hexagonal close packed crystals, for example, have three main families - basal, prismatic, and pyramidal - with different values for the critical resolved shear stress.

Slip Systems and Resolved Shear Stress

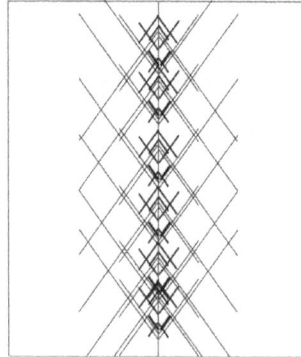

Slip systems activated near grain boundary to ensure compatibility.

In crystalline metals, slip occurs in specific directions on crystallographic planes, and each combination of slip direction and slip plane will have its own Schmid factor. As an example, for a face-centered cubic (FCC) system the primary slip planes and directions belong to the {111} and <110> families. The Schmid Factor for the $(\bar{1}11)$ plane in the [101] direction for an applied tensile load in the [123] direction can be calculated as:

$$m_{[1\,0\,1](\bar{1}11)} = \frac{(-1*1)+(1*2)+(1*3)}{((-1)^2+1^2+1^2)^{1/2}(1^2+2^2+3^2)^{1/2}} * \frac{(1*1)+(0*2)+(1*3)}{(1^2+0^2+1^2)^{1/2}(1^2+2^2+3^2)^{1/2}} = \frac{16}{14\sqrt{6}}$$

In a single crystal sample, the macroscopic yield stress will be determined by the Schmid factor of the single grain. Thus, in general, different yield strengths will be observed for applied stresses along different crystallographic directions. In polycrystalline specimens, the yield strength of each grain is different depending on its maximum Schmid factor, which indicates the operational slip system(s). The macroscopically observed yield stress will be related to the material's CRSS by an average Schmid factor, which is roughly 1/3.06 for FCC and 1/2.75 for body-centered cubic (BCC) structures.

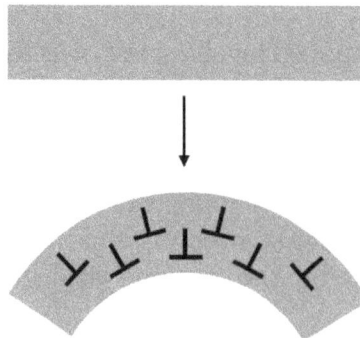

Geometrically necessary dislocations for bending of a bar of material.

The onset of plasticity in polycrystals is influenced by the number of available slip systems to accommodate incompatibilities at the grain boundaries. In the case of two adjacent, randomly oriented grains, one grain will have a larger Schmid factor and thus a smaller yield stress. Under load, this "weaker" grain will yield prior to the "stronger" grain, and as it deforms a stress concentration will build up in the stronger grain near the boundary between them. This stress concentration will activate dislocation motion in the available glide planes. These dislocations are geometrically necessary to ensure that the strain in each grain is equivalent at the grain boundary, so that the compatibility criteria are satisfied. Taylor showed that a minimum of five active slip systems are required to accommodate an arbitrary deformation. In crystal structures with fewer than 5 active slip systems, such as hexagonal close-packed (HCP) metals, the specimen will exhibit brittle failure instead of plastic deformation.

Slip Systems in Crystalline Metals		
Crystal Structure	Primary Slip System	Number of Independent Systems
Face-centered cubic (FCC)	{111}<1-10>	5
Body-centered cubic (BCC)	{110}<-111>	5
Hexagonal close-packed (HCP)	{0001}<11-20>	2

Effects of Temperature and Solid Solution Strengthening

At lower temperatures, more energy (i.e. - larger applied stress) is required to activate some slip systems. This is particularly evident in BCC materials, in which not all 5 independent slip systems are thermally activated at temperatures below the ductile-to-brittle transition temperature, or DBTT, so BCC specimens therefore become brittle. In general, the CRSS increases as the homologous temperature decreases because it becomes energetically more costly to activate the slip systems, although this effect is much less pronounced in FCC.

Solid solution strengthening also increases the CRSS compared to a pure single component material because the solute atoms distort the lattice, preventing the dislocation motion necessary for plasticity. With dislocation motion inhibited, it becomes harder to activate the necessary 5 independent slip systems, so the material becomes stronger and more brittle.

Crystal Defects in Metals

The detailed information of sequenced arrangement of atoms, ions or molecules in a crystalline material is called crystal structure. The crystal structure, however, feature many defects in them such as vacancy defects. When the atom is missing from a lattice side, it is called vacancy defect. The aspects elucidated in this section are of vital importance, and provide a better understanding of physical metallurgy.

Crystal Structure

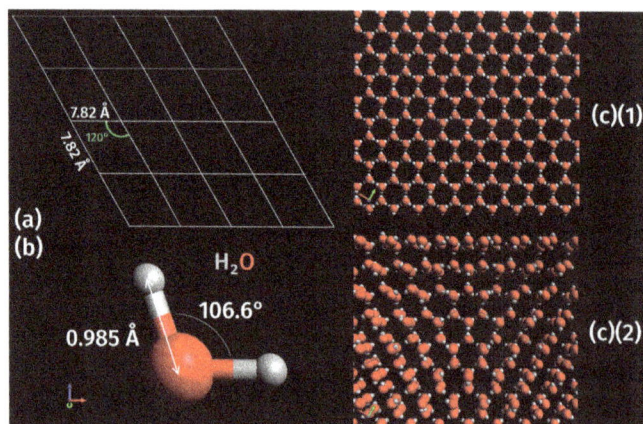

The (3-D) crystal structure of H_2O ice Ih (c) consists of bases of H_2O ice molecules (b) located on lattice points within the (2-D) hexagonal space lattice (a). The values for the H–O–H angle and O–H distance have come from *Physics of Ice* with uncertainties of ±1.5° and ±0.005 Å, respectively. The white box in (c) is the unit cell defined by Bernal and Fowler

In crystallography, crystal structure is a description of the ordered arrangement of atoms, ions or molecules in a crystalline material. Ordered structures occur from the intrinsic nature of the constituent particles to form symmetric patterns that repeat along the principal directions of three-dimensional space in matter.

The smallest group of particles in the material that constitutes the repeating pattern is the unit cell of the structure. The unit cell completely defines the symmetry and structure of the entire crystal lattice, which is built up by repetitive translation of the unit cell along its principal axes. The repeating patterns are said to be located at the points of the Bravais lattice.

The lengths of the principal axes, or edges, of the unit cell and the angles between them are the lattice constants, also called *lattice parameters*. The symmetry properties of the crystal are described by the concept of space groups. All possible symmetric arrangements of particles in three-dimensional space may be described by the 230 space groups.

The crystal structure and symmetry play a critical role in determining many physical properties, such as cleavage, electronic band structure, and optical transparency.

Unit Cell

The crystal structure of a material (the arrangement of atoms within a given type of crystal) can be described in terms of its unit cell. The unit cell is a box containing one or more atoms arranged in three dimensions. The unit cells stacked in three-dimensional space describe the bulk arrangement of atoms of the crystal. The unit cell is represented in terms of its lattice parameters, which are the lengths of the cell edges (a, b and c) and the angles between them (alpha, beta and gamma), while the positions of the atoms inside the unit cell are described by the set of atomic positions (x_i, y_i, z_i) measured from a reference lattice point. Commonly, atomic positions are represented in terms of fractional coordinates, relative to the unit cell lengths.

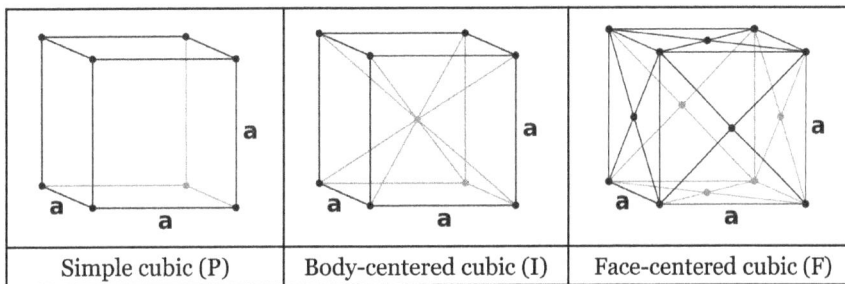

| Simple cubic (P) | Body-centered cubic (I) | Face-centered cubic (F) |

The atom positions within the unit cell can be calculated through application of symmetry operations to the asymmetric unit. The asymmetric unit refers to the smallest possible occupation of space within the unit cell. This does not, however imply that the entirety of the asymmetric unit must lie within the boundaries of the unit cell. Symmetric transformations of atom positions are calculated from the space group of the crystal structure, and this is usually a black box operation performed by computer programs. However, manual calculation of the atomic positions within the unit cell can be performed from the asymmetric unit, through the application of the symmetry operators described within the *International Tables for Crystallography: Volume A*.

Miller Indices

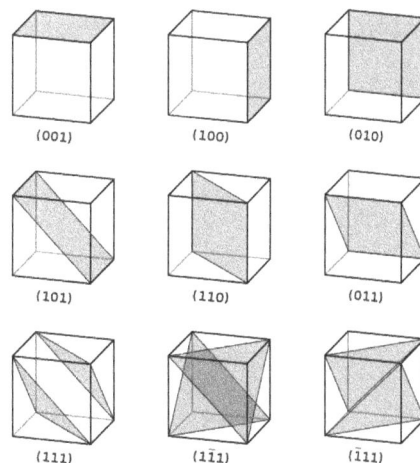

Planes with different Miller indices in cubic crystals

Vectors and planes in a crystal lattice are described by the three-value Miller index notation. It uses the indices ℓ, m, and n as directional parameters, which are separated by 90°, and are thus orthogonal.

By definition, the syntax (ℓmn) denotes a plane that intercepts the three points a_1/ℓ, a_2/m, and a_3/n, or some multiple thereof. That is, the Miller indices are proportional to the inverses of the intercepts of the plane with the unit cell (in the basis of the lattice vectors). If one or more of the indices is zero, it means that the planes do not intersect that axis (i.e., the intercept is "at infinity"). A plane containing a coordinate axis is translated so that it no longer contains that axis before its Miller indices are determined. The Miller indices for a plane are integers with no common factors. Negative indices are indicated with horizontal bars, as in (123). In an orthogonal coordinate system for a cubic cell, the Miller indices of a plane are the Cartesian components of a vector normal to the plane.

Considering only (ℓmn) planes intersecting one or more lattice points (the *lattice planes*), the distance d between adjacent lattice planes is related to the (shortest) reciprocal lattice vector orthogonal to the planes by the formula

$$d = \frac{2\pi}{|\mathbf{g}_{\ell mn}|}$$

Planes and Directions

The crystallographic directions are geometric lines linking nodes (atoms, ions or molecules) of a crystal. Likewise, the crystallographic planes are geometric *planes* linking nodes. Some directions and planes have a higher density of nodes. These high density planes have an influence on the behavior of the crystal as follows:

*   Optical properties: Refractive index is directly related to density (or periodic density fluctuations).

*   Adsorption and reactivity: Physical adsorption and chemical reactions occur at or near surface atoms or molecules. These phenomena are thus sensitive to the density of nodes.

*   Surface tension: The condensation of a material means that the atoms, ions or molecules are more stable if they are surrounded by other similar species. The surface tension of an interface thus varies according to the density on the surface.

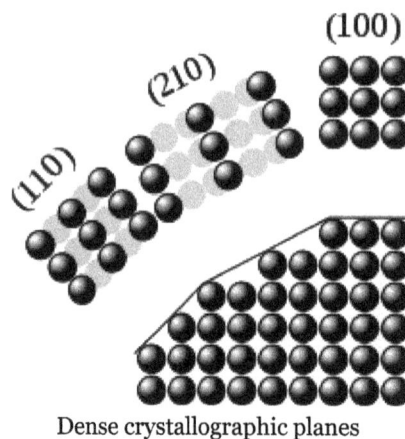

Dense crystallographic planes

- Microstructural defects: Pores and crystallites tend to have straight grain boundaries following higher density planes.

- Cleavage: This typically occurs preferentially parallel to higher density planes.

- Plastic deformation: Dislocation glide occurs preferentially parallel to higher density planes. The perturbation carried by the dislocation (Burgers vector) is along a dense direction. The shift of one node in a more dense direction requires a lesser distortion of the crystal lattice.

Some directions and planes are defined by symmetry of the crystal system. In monoclinic, rhombohedral, tetragonal, and trigonal/hexagonal systems there is one unique axis (sometimes called the principal axis) which has higher rotational symmetry than the other two axes. The basal plane is the plane perpendicular to the principal axis in these crystal systems. For triclinic, orthorhombic, and cubic crystal systems the axis designation is arbitrary and there is no principal axis.

Cubic Structures

For the special case of simple cubic crystals, the lattice vectors are orthogonal and of equal length (usually denoted a); similarly for the reciprocal lattice. So, in this common case, the Miller indices (ℓmn) and [ℓmn] both simply denote normals/directions in Cartesian coordinates. For cubic crystals with lattice constant a, the spacing d between adjacent (ℓmn) lattice planes is:

$$d_{\ell mn} = \frac{a}{\sqrt{\ell^2 + m^2 + n^2}}$$

Because of the symmetry of cubic crystals, it is possible to change the place and sign of the integers and have equivalent directions and planes:

- Coordinates in *angle brackets* such as ⟨100⟩ denote a *family* of directions that are equivalent due to symmetry operations, such as [100], [010], [001] or the negative of any of those directions.

- Coordinates in *curly brackets* or *braces* such as {100} denote a family of plane normals that are equivalent due to symmetry operations, much the way angle brackets denote a family of directions.

For face-centered cubic (fcc) and body-centered cubic (bcc) lattices, the primitive lattice vectors are not orthogonal. However, in these cases the Miller indices are conventionally defined relative to the lattice vectors of the cubic supercell and hence are again simply the Cartesian directions.

Interplanar Spacing

The spacing d between adjacent (hkl) lattice planes is given by:

- Cubic:

$$\frac{1}{d^2} = \frac{h^2 + k^2 + l^2}{a^2}$$

- Tetragonal:

$$\frac{1}{d^2} = \frac{h^2 + k^2}{a^2} + \frac{l^2}{c^2}$$

- Hexagonal:

$$\frac{1}{d^2} = \frac{4}{3}\left(\frac{h^2 + hk + k^2}{a^2}\right) + \frac{l^2}{c^2}$$

- Rhombohedral:

$$\frac{1}{d^2} = \frac{(h^2 + k^2 + l^2)\sin^2\alpha + 2(hk + kl + hl)(\cos^2\alpha - \cos\alpha)}{a^2(1 - 3\cos^2\alpha + 2\cos^3\alpha)}$$

- Orthorhombic:

$$\frac{1}{d^2} = \frac{h^2}{a^2} + \frac{k^2}{b^2} + \frac{l^2}{c^2}$$

- Monoclinic:

$$\frac{1}{d^2} = \left(\frac{h^2}{a^2} + \frac{k^2\sin^2\beta}{b^2} + \frac{l^2}{c^2} - \frac{2hl\cos\beta}{ac}\right)\csc^2\beta$$

- Triclinic:

$$\frac{1}{d^2} = \frac{\dfrac{h^2}{a^2}\sin^2\alpha + \dfrac{k^2}{b^2}\sin^2\beta + \dfrac{l^2}{c^2}\sin^2\gamma + \dfrac{2kl}{bc}\cos\alpha + \dfrac{2hl}{ac}\cos\beta + \dfrac{2hk}{ab}\cos\gamma}{1 - \cos^2\alpha - \cos^2\beta - \cos^2\gamma + 2\cos\alpha\cos\beta\cos\gamma}$$

Classification by Symmetry

The defining property of a crystal is its inherent symmetry, by which we mean that under certain 'operations' the crystal remains unchanged. All crystals have translational symmetry in three directions, but some have other symmetry elements as well. For example, rotating the crystal 180° about a certain axis may result in an atomic configuration that is identical to the original configuration. The crystal is then said to have a twofold rotational symmetry about this axis. In addition to rotational symmetries like this, a crystal may have symmetries in the form of mirror planes and translational symmetries, and also the so-called "compound symmetries," which are a combination of translation and rotation/mirror symmetries. A full classification of a crystal is achieved when all of these inherent symmetries of the crystal are identified.

Lattice Systems

These lattice systems are a grouping of crystal structures according to the axial system used to describe their lattice. Each lattice system consists of a set of three axes in a particular geometric arrangement. There are seven lattice systems. They are similar to but not quite the same as the seven crystal systems.

Crystal family	Lattice system	Schönflies	14 Bravais Lattices			
			Primitive	Base-centered	Body-centered	Face-centered
triclinic	C_i					
monoclinic	C_{2h}		$\beta \neq 90°$ $a \neq c$	$\beta \neq 90°$ $a \neq c$		
orthorhombic	D_{2h}		$a \neq b \neq c$	$a \neq b \neq c$	$a \neq b \neq c$	$a \neq b \neq c$
tetragonal	D_{4h}		$a \neq c$		$a \neq c$	

The simplest and most symmetric, the cubic (or isometric) system, has the symmetry of a cube, that is, it exhibits four threefold rotational axes oriented at 109.5° (the tetrahedral angle) with respect to each other. These threefold axes lie along the body diagonals of the cube. The other six lattice systems, are hexagonal, tetragonal, rhombohedral (often confused with the trigonal crystal system), orthorhombic, monoclinic and triclinic.

Bravais Lattices

Bravais lattices, also referred to as *space lattices*, describe the geometric arrangement of the lattice points, and therefore the translational symmetry of the crystal. The three dimensions of space afford 14 distinct Bravais lattices describing the translational symmetry. All crystalline materials recognized today, not including quasicrystals, fit in one of these arrangements. The fourteen three-dimensional lattices, classified by lattice system.

The crystal structure consists of the same group of atoms, the *basis*, positioned around each and every lattice point. This group of atoms therefore repeats indefinitely in three dimensions according to the arrangement of one of the Bravais lattices. The characteristic rotation and mirror symmetries of the unit cell is described by its crystallographic point group.

Crystal Systems

A crystal system is a set of point groups in which the point groups themselves and their corresponding space groups are assigned to a lattice system. Of the 32 point groups that exist in three dimensions, most are assigned to only one lattice system, in which case the crystal system and lattice system both have the same name. However, five point groups are assigned to two lattice systems, rhombohedral and hexagonal, because both lattice systems exhibit threefold rotational symmetry. In total there are seven crystal systems: triclinic, monoclinic, orthorhombic, tetragonal, trigonal, hexagonal, and cubic.

Point Groups

The crystallographic point group or *crystal class* is the mathematical group comprising the symmetry operations that leave at least one point unmoved and that leave the appearance of the crystal structure unchanged. These symmetry operations include

- *Reflection*, which reflects the structure across a *reflection plane*

- *Rotation*, which rotates the structure a specified portion of a circle about a *rotation axis*

- *Inversion*, which changes the sign of the coordinate of each point with respect to a *center of symmetry* or *inversion point*

- *Improper rotation*, which consists of a rotation about an axis followed by an inversion.

Rotation axes (proper and improper), reflection planes, and centers of symmetry are collectively called *symmetry elements*. There are 32 possible crystal classes. Each one can be classified into one of the seven crystal systems.

Space Groups

In addition to the operations of the point group, the space group of the crystal structure contains translational symmetry operations. These include:

- Pure *translations*, which move a point along a vector

- *Screw axes*, which rotate a point around an axis while translating parallel to the axis.

- *Glide planes*, which reflect a point through a plane while translating it parallel to the plane.

There are 230 distinct space groups.

Atomic Coordination

By considering the arrangement of atoms relative to each other, their coordination numbers (or number of nearest neighbors), interatomic distances, types of bonding, etc., it is possible to form a general view of the structures and alternative ways of visualizing them.

Close Packing

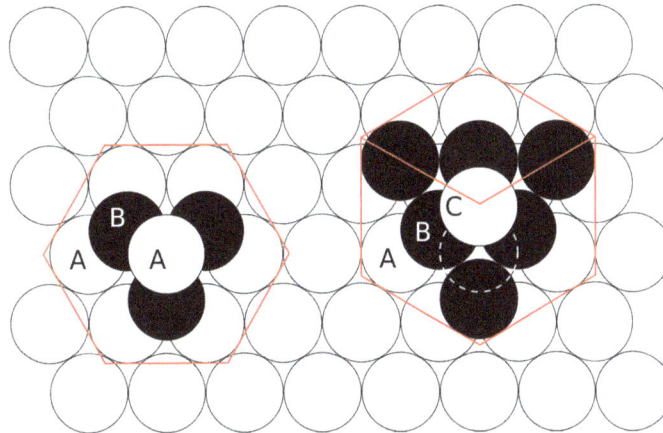

The hcp lattice (left) and the fcc lattice (right)

The principles involved can be understood by considering the most efficient way of packing together equal-sized spheres and stacking close-packed atomic planes in three dimensions. For example, if plane A lies beneath plane B, there are two possible ways of placing an additional atom on top of layer B. If an additional layer was placed directly over plane A, this would give rise to the following series:

...ABABABAB...

This arrangement of atoms in a crystal structure is known as hexagonal close packing (hcp).

If, however, all three planes are staggered relative to each other and it is not until the fourth layer is positioned directly over plane A that the sequence is repeated, then the following sequence arises:

...ABCABCABC...

This type of structural arrangement is known as cubic close packing (ccp).

The unit cell of a ccp arrangement of atoms is the face-centered cubic (fcc) unit cell. This is not immediately obvious as the closely packed layers are parallel to the {111} planes of the fcc unit cell. There are four different orientations of the close-packed layers.

The packing efficiency can be worked out by calculating the total volume of the spheres and dividing by the volume of the cell as follows:

$$\frac{4 \times \frac{4}{3}\pi r^3}{16\sqrt{2}r^3} = \frac{\pi}{3\sqrt{2}} = 0.7405...$$

The 74% packing efficiency is the maximum density possible in unit cells constructed of spheres of only one size. Most crystalline forms of metallic elements are hcp, fcc, or bcc (body-centered cubic). The coordination number of atoms in hcp and fcc structures is 12 and its atomic packing factor (APF) is the number mentioned above, 0.74. This can be compared to the APF of a bcc structure, which is 0.68.

Grain Boundaries

Grain boundaries are interfaces where crystals of different orientations meet. A grain boundary is a single-phase interface, with crystals on each side of the boundary being identical except in orientation. The term "crystallite boundary" is sometimes, though rarely, used. Grain boundary areas contain those atoms that have been perturbed from their original lattice sites, dislocations, and impurities that have migrated to the lower energy grain boundary.

Treating a grain boundary geometrically as an interface of a single crystal cut into two parts, one of which is rotated, we see that there are five variables required to define a grain boundary. The first two numbers come from the unit vector that specifies a rotation axis. The third number designates the angle of rotation of the grain. The final two numbers specify the plane of the grain boundary (or a unit vector that is normal to this plane).

Grain boundaries disrupt the motion of dislocations through a material, so reducing crystallite size is a common way to improve strength, as described by the Hall–Petch relationship. Since grain boundaries are defects in the crystal structure they tend to decrease the electrical and thermal conductivity of the material. The high interfacial energy and relatively weak bonding in most grain boundaries often makes them preferred sites for the onset of corrosion and for the precipitation of new phases from the solid. They are also important to many of the mechanisms of creep.

Grain boundaries are in general only a few nanometers wide. In common materials, crystallites are large enough that grain boundaries account for a small fraction of the material. However, very small grain sizes are achievable. In nanocrystalline solids, grain boundaries become a significant volume fraction of the material, with profound effects on such properties as diffusion and plasticity. In the limit of small crystallites, as the volume fraction of grain boundaries approaches 100%, the material ceases to have any crystalline character, and thus becomes an amorphous solid.

Defects and Impurities

Real crystals feature defects or irregularities in the ideal arrangements described above and it is these defects that critically determine many of the electrical and mechanical properties of real materials. When one atom substitutes for one of the principal atomic components within the crystal structure, alteration in the electrical and thermal properties of the material may ensue. Impurities may also manifest as spin impurities in certain materials. Research on magnetic impurities demonstrates that substantial alteration of certain properties such as specific heat may be affected by small concentrations of an impurity, as for example impurities in semiconducting ferromagnetic alloys may lead to different properties as first predicted in the late 1960s. Dislocations in the crystal lattice allow shear at lower stress than that needed for a perfect crystal structure.

Prediction of Structure

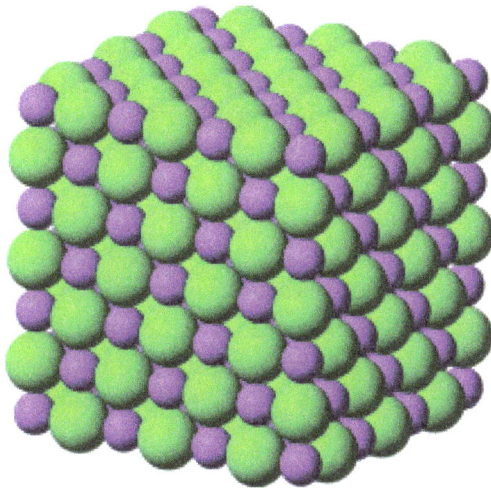

Crystal structure of sodium chloride (table salt)

The difficulty of predicting stable crystal structures based on the knowledge of only the chemical composition has long been a stumbling block on the way to fully computational materials design. Now, with more powerful algorithms and high-performance computing, structures of medium complexity can be predicted using such approaches as evolutionary algorithms, random sampling, or metadynamics.

The crystal structures of simple ionic solids (e.g., NaCl or table salt) have long been rationalized in terms of Pauling's rules, first set out in 1929 by Linus Pauling, referred to by many since as the "father of the chemical bond". Pauling also considered the nature of the interatomic forces in metals, and concluded that about half of the five d-orbitals in the transition metals are involved in bonding, with the remaining nonbonding d-orbitals being responsible for the magnetic properties. He, therefore, was able to correlate the number of d-orbitals in bond formation with the bond length as well as many of the physical properties of the substance. He subsequently introduced the metallic orbital, an extra orbital necessary to permit uninhibited resonance of valence bonds among various electronic structures.

In the resonating valence bond theory, the factors that determine the choice of one from among alternative crystal structures of a metal or intermetallic compound revolve around the energy of resonance of bonds among interatomic positions. It is clear that some modes of resonance would make larger contributions (be more mechanically stable than others), and that in particular a simple ratio of number of bonds to number of positions would be exceptional. The resulting principle is that a special stability is associated with the simplest ratios or "bond numbers": $\frac{1}{2}$, $\frac{1}{3}$, $\frac{2}{3}$, $\frac{1}{4}$, $\frac{3}{4}$, etc. The choice of structure and the value of the axial ratio (which determines the relative bond lengths) are thus a result of the effort of an atom to use its valency in the formation of stable bonds with simple fractional bond numbers.

After postulating a direct correlation between electron concentration and crystal structure in beta-phase alloys, Hume-Rothery analyzed the trends in melting points, compressibilities and bond lengths as a function of group number in the periodic table in order to establish a system of valencies of the transition elements in the metallic state. This treatment thus emphasized the increasing bond strength as a function of group number. The operation of directional forces were emphasized in one article on the relation between bond hybrids and the metallic structures. The resulting correlation between electronic and crystalline structures is summarized by a single parameter, the weight of the d-electrons per hybridized metallic orbital. The "d-weight" calculates out to 0.5, 0.7 and 0.9 for the fcc, hcp and bcc structures respectively. The relationship between d-electrons and crystal structure thus becomes apparent.

In crystal structure predictions/simulations, the periodicity is usually applied, since the system is imagined as unlimited big in all directions. Starting from a triclinic structure with no further symmetry property assumed, the system may be driven to show some additional symmetry properties by applying Newton's Second Law on particles in the unit cell and a recently developed dynamical equation for the system period vectors (lattice parameters including angles), even if the system is subject to external stress.

Polymorphism

Quartz is one of the several thermodynamically stable crystalline forms of silica, SiO_2.
The most important forms of silica include: α-quartz, β-quartz, tridymite, cristobalite, coesite, and stishovite.

Polymorphism is the occurrence of multiple crystalline forms of a material. It is found in many crystalline materials including polymers, minerals, and metals. According to Gibbs' rules of phase equilibria, these unique crystalline phases are dependent on intensive variables such as pressure and temperature. Polymorphism is related to allotropy, which refers to elemental solids. The complete morphology of a material is described by polymorphism and other variables such as crystal habit, amorphous fraction or crystallographic defects. Polymorphs have different stabilities and may spontaneously convert from a metastable form (or thermodynamically unstable form) to the stable form at a particular temperature. They also exhibit different melting points, solubilities, and X-ray diffraction patterns.

One good example of this is the quartz form of silicon dioxide, or SiO_2. In the vast majority of silicates, the Si atom shows tetrahedral coordination by 4 oxygens. All but one of the crystalline forms involve tetrahedral $\{SiO_4\}$ units linked together by shared vertices in different arrangements. In different minerals the tetrahedra show different degrees of networking and polymerization. For example, they occur singly, joined together in pairs, in larger finite clusters including rings, in chains, double chains, sheets, and three-dimensional frameworks. The minerals are classified into groups based on these structures. In each of its 7 thermodynamically stable crystalline forms or polymorphs of crystalline quartz, only 2 out of 4 of each the edges of the $\{SiO_4\}$ tetrahedra are shared with others, yielding the net chemical formula for silica: SiO_2.

Another example is elemental tin (Sn), which is malleable near ambient temperatures but is brittle when cooled. This change in mechanical properties due to existence of its two major allotropes, α- and β-tin. The two allotropes that are encountered at normal pressure and temperature, α-tin and β-tin, are more commonly known as *gray tin* and *white tin* respectively. Two more allotropes, γ and σ, exist at temperatures above 161 °C and pressures above several GPa. White tin is metallic, and is the stable crystalline form at or above room temperature. Below 13.2 °C, tin exists in the gray form, which has a diamond cubic crystal structure, similar to diamond, silicon or germanium. Gray tin has no metallic properties at all, is a dull gray powdery material, and has few uses, other than a few specialized semiconductor applications. Although the $\alpha-\beta$ transformation temperature of tin is nominally 13.2 °C, impurities (e.g. Al, Zn, etc.) lower the transition temperature well below 0 °C, and upon addition of Sb or Bi the transformation may not occur at all.

Physical Properties

Twenty of the 32 crystal classes are piezoelectric, and crystals belonging to one of these classes (point groups) display piezoelectricity. All piezoelectric classes lack a center of symmetry. Any material develops a dielectric polarization when an electric field is applied, but a substance that has such a natural charge separation even in the absence of a field is called a polar material. Whether or not a material is polar is determined solely by its crystal structure. Only 10 of the 32 point groups are polar. All polar crystals are pyroelectric, so the 10 polar crystal classes are sometimes referred to as the pyroelectric classes.

There are a few crystal structures, notably the perovskite structure, which exhibit ferroelectric behavior. This is analogous to ferromagnetism, in that, in the absence of an electric field during production, the ferroelectric crystal does not exhibit a polarization. Upon the application of an electric field of sufficient magnitude, the crystal becomes permanently polarized. This polarization can be reversed by a sufficiently large counter-charge, in the same way that a ferromagnet can be reversed. However, although they are called ferroelectrics, the effect is due to the crystal structure (not the presence of a ferrous metal).

Crystallographic Defect

Crystalline solids exhibit a periodic crystal structure. The positions of atoms or molecules occur on repeating fixed distances, determined by the unit cell parameters. However, the arrangement of atoms or molecules in most crystalline materials is not perfect. The regular patterns are interrupted by crystallographic defects.

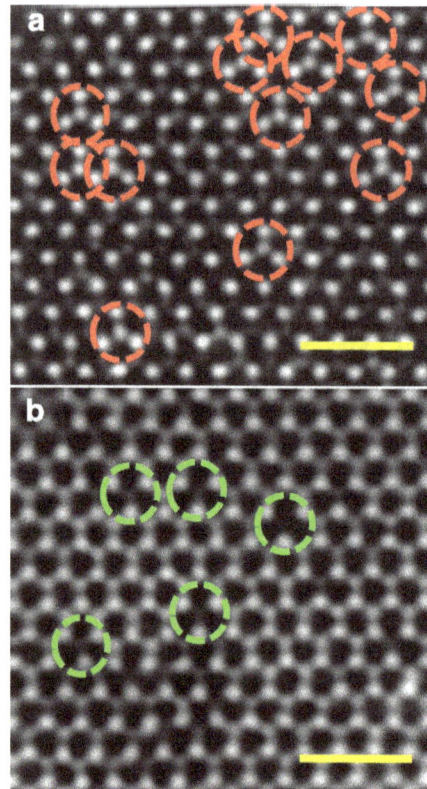

Electron microscopy of antisites (a, Mo substitutes for S) and vacancies (b, missing S atoms)
in a monolayer of molybdenum disulfide. Scale bar: 1 nm.

Point Defects

Point defects are defects that occur only at or around a single lattice point. They are not extended
in space in any dimension. Strict limits for how small a point defect is are generally not defined
explicitly. However, these defects typically involve at most a few extra or missing atoms. Larger
defects in an ordered structure are usually considered dislocation loops. For historical reasons,
many point defects, especially in ionic crystals, are called *centers*: for example a vacancy in many
ionic solids is called a luminescence center, a color center, or F-center. These dislocations permit
ionic transport through crystals leading to electrochemical reactions. These are frequently specified using Kröger–Vink Notation.

- Vacancy defects are lattice sites which would be occupied in a perfect crystal, but are vacant. If a neighboring atom moves to occupy the vacant site, the vacancy moves in the
  opposite direction to the site which used to be occupied by the moving atom. The stability
  of the surrounding crystal structure guarantees that the neighboring atoms will not simply
  collapse around the vacancy. In some materials, neighboring atoms actually move away
  from a vacancy, because they experience attraction from atoms in the surroundings. A vacancy (or pair of vacancies in an ionic solid) is sometimes called a Schottky defect.

- Interstitial defects are atoms that occupy a site in the crystal structure at which there is
  usually not an atom. They are generally high energy configurations. Small atoms in some
  crystals can occupy interstices without high energy, such as hydrogen in palladium.

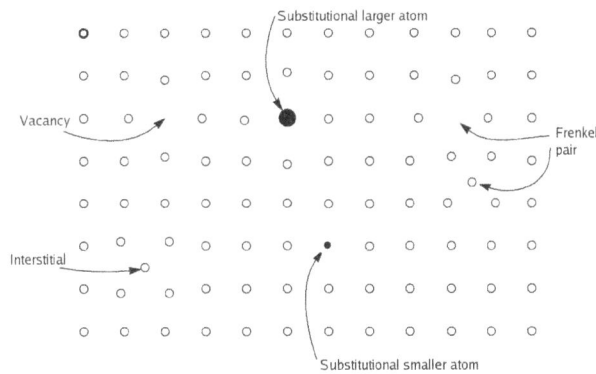

Schematic illustration of some simple point defect types in a monatomic solid

- A nearby pair of a vacancy and an interstitial is often called a Frenkel defect or Frenkel pair. This is caused when an ion moves into an interstitial site and creates a vacancy.

- Due to fundamental limitations of material purification methods, materials are never 100% pure, which by definition induces defects in crystal structure. In the case of an impurity, the atom is often incorporated at a regular atomic site in the crystal structure. This is neither a vacant site nor is the atom on an interstitial site and it is called a *substitutional* defect. The atom is not supposed to be anywhere in the crystal, and is thus an impurity. In some cases where the radius of the substitutional atom (ion) is substantially smaller than that of the atom (ion) it is replacing, its equilibrium position can be shifted away from the lattice site. These types of substitutional defects are often referred to as off-center ions. There are two different types of substitutional defects: Isovalent substitution and aliovalent substitution. Isovalent substitution is where the ion that is substituting the original ion is of the same oxidation state as the ion it is replacing. Aliovalent substitution is where the ion that is substituting the original ion is of a different oxidation state than the ion it is replacing. Aliovalent substitutions change the overall charge within the ionic compound, but the ionic compound must be neutral. Therefore, a charge compensation mechanism is required. Hence either one of the metals is partially or fully oxidised or reduced, or ion vacancies are created.

- Antisite defects occur in an ordered alloy or compound when atoms of different type exchange positions. For example, some alloys have a regular structure in which every other atom is a different species; for illustration assume that type A atoms sit on the corners of a cubic lattice, and type B atoms sit in the center of the cubes. If one cube has an A atom at its center, the atom is on a site usually occupied by a B atom, and is thus an antisite defect. This is neither a vacancy nor an interstitial, nor an impurity.

- Topological defects are regions in a crystal where the normal chemical bonding environment is topologically different from the surroundings. For instance, in a perfect sheet of graphite (graphene) all atoms are in rings containing six atoms. If the sheet contains regions where the number of atoms in a ring is different from six, while the total number of atoms remains the same, a topological defect has formed. An example is the Stone Wales defect in nanotubes, which consists of two adjacent 5-membered and two 7-membered atom rings.

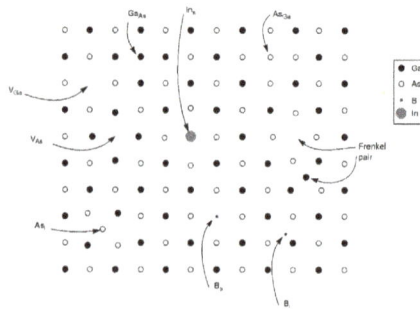

Schematic illustration of defects in a compound solid, using GaAs as an example.

- Also amorphous solids may contain defects. These are naturally somewhat hard to define, but sometimes their nature can be quite easily understood. For instance, in ideally bonded amorphous silica all Si atoms have 4 bonds to O atoms and all O atoms have 2 bonds to Si atom. Thus e.g. an O atom with only one Si bond (a dangling bond) can be considered a defect in silica. Moreover, defects can also be defined in amorphous solids based on empty or densely packed local atomic neighbourhoods, and the properties of such 'defects' can be shown to be similar to normal vacancies and interstitials in crystals.

- Complexes can form between different kinds of point defects. For example, if a vacancy encounters an impurity, the two may bind together if the impurity is too large for the lattice. Interstitials can form 'split interstitial' or 'dumbbell' structures where two atoms effectively share an atomic site, resulting in neither atom actually occupying the site.

Line Defects

Line defects can be described by gauge theories.

Dislocations are linear defects, around which (the dislocation line) some of the atoms of the crystal lattice are misaligned. There are two basic types of dislocations, the *edge* dislocation and the *screw* dislocation. "Mixed" dislocations, combining aspects of both types, are also common.

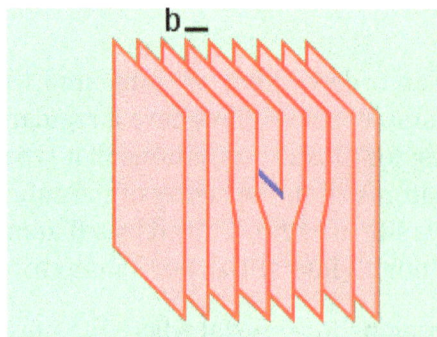

An *edge dislocation* is shown. The dislocation line is presented in blue, the Burgers vector b in black.

Edge dislocations are caused by the termination of a plane of atoms in the middle of a crystal. In such a case, the adjacent planes are not straight, but instead bend around the edge of the terminating plane so that the crystal structure is perfectly ordered on either side. The analogy with a stack of paper is apt: if a half a piece of paper is inserted in a stack of paper, the defect in the stack is only noticeable at the edge of the half sheet.

The screw dislocation is more difficult to visualise, but basically comprises a structure in which a helical path is traced around the linear defect (dislocation line) by the atomic planes of atoms in the crystal lattice.

The presence of dislocation results in lattice strain (distortion). The direction and magnitude of such distortion is expressed in terms of a Burgers vector (b). For an edge type, b is perpendicular to the dislocation line, whereas in the cases of the screw type it is parallel. In metallic materials, b is aligned with close-packed crystallographic directions and its magnitude is equivalent to one interatomic spacing.

Dislocations can move if the atoms from one of the surrounding planes break their bonds and rebond with the atoms at the terminating edge.

It is the presence of dislocations and their ability to readily move (and interact) under the influence of stresses induced by external loads that leads to the characteristic malleability of metallic materials.

Dislocations can be observed using transmission electron microscopy, field ion microscopy and atom probe techniques. Deep level transient spectroscopy has been used for studying the electrical activity of dislocations in semiconductors, mainly silicon.

Disclinations are line defects corresponding to "adding" or "subtracting" an angle around a line. Basically, this means that if you track the crystal orientation around the line defect, you get a rotation. Usually, they were thought to play a role only in liquid crystals, but recent developments suggest that they might have a role also in solid materials, e.g. leading to the self-healing of cracks.

Planar Defects

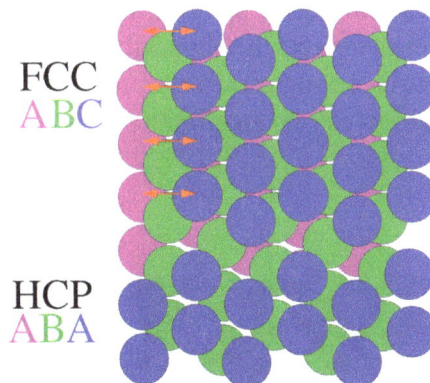

Origin of stacking faults: Different stacking sequences of close-packed crystals

- Grain boundaries occur where the crystallographic direction of the lattice abruptly changes. This usually occurs when two crystals begin growing separately and then meet.

- Antiphase boundaries occur in ordered alloys: in this case, the crystallographic direction remains the same, but each side of the boundary has an opposite phase: For example, if the ordering is usually ABABABAB (hexagonal close-packed crystal), an antiphase boundary takes the form of ABABBABA.

- Stacking faults occur in a number of crystal structures, but the common example is in close-packed structures. They are formed by a local deviation of the stacking sequence of layers in a crystal. An example would be the ABABCABAB stacking sequence.

- A twin boundary is a defect that introduces a plane of mirror symmetry in the ordering of a crystal. For example, in cubic close-packed crystals, the stacking sequence of a twin boundary would be ABCABCBACBA.

- On surfaces of single crystals, steps between atomically flat terraces can also be regarded as planar defects. It has been shown that such defects and their geometry have significant influence on the adsorption of organic molecules

Bulk Defects

- Three-dimensional macroscopic or bulk defects, such as pores, cracks, or inclusions.

- Voids — small regions where there are no atoms, and which can be thought of as clusters of vacancies.

- Impurities can cluster together to form small regions of a different phase. These are often called precipitates.

Mathematical Classification Methods

A successful mathematical classification method for physical lattice defects, which works not only with the theory of dislocations and other defects in crystals but also, e.g., for disclinations in liquid crystals and for excitations in superfluid He, is the topological homotopy theory.

Computer Simulation Methods

Density functional theory, classical molecular dynamics and kinetic Monte Carlo simulations are widely used to study the properties of defects in solids with computer simulations. Simulating jamming of hard spheres of different sizes and/or in containers with non-commeasurable sizes using the Lubachevsky–Stillinger algorithm can be an effective technique for demonstrating some types of crystallographic defects.

Vacancy Defect

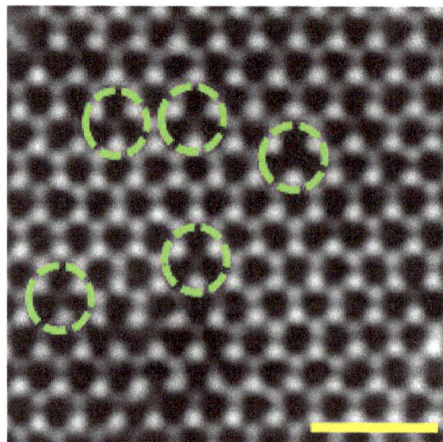

Electron microscopy of sulfur vacancies in a monolayer of molybdenum disulfide. Right circle points to a divacancy, i.e., sulfur atoms are missing both above and below the Mo layer. Other circles are single vacancies, i.e., sulfur atoms are missing only above or below the Mo layer. Scale bar: 1 nm.

In crystallography, a vacancy is a type of point defect in a crystal. Crystals inherently possess imperfections, sometimes referred to as crystalline defects. A defect in which an atom is missing from one of the lattice sites is known as a "vacancy" defect. It is also known as a Schottky defect, although in ionic crystals the concepts are not identical.

Vacancies occur naturally in all crystalline materials. At any given temperature, up to the melting point of the material, there is an equilibrium concentration (ratio of vacant lattice sites to those containing atoms). At the melting point of some metals the ratio can be approximately 1:1000. This temperature dependence can be modeled by

$$N_v = N\exp(-Q_v / k_B T)$$

where N_v is the vacancy concentration, Q_v is the energy required for vacancy formation, k_B is the Boltzmann constant, T it the absolute temperature, and N is the concentration of atomic sites i.e.

$$N = \rho N_A / A$$

where ρ is density, N_A Avogadro constant, and A the atomic mass.

It is the simplest point defect. In this system, an atom is missing from its regular atomic site. Vacancies are formed during solidification due to vibration of atoms, local rearrangement of atoms, plastic deformation and ionic bombardments.

The creation of a vacancy can be simply modeled by considering the energy required to break the bonds between an atom inside the crystal and its nearest neighbor atoms. Once that atom is removed from the lattice site, it is put back on the surface of the crystal and some energy is retrieved because new bonds are established with other atoms on the surface. However, there is a net input of energy because there are fewer bonds between surface atoms than between atoms in the interior of the crystal.

Dislocation

In materials science, a dislocation or Taylor's dislocation is a crystallographic defect or irregularity within a crystal structure. The presence of dislocations strongly influences many of the properties of materials.

The theory describing the elastic fields of the defects was originally developed by Vito Volterra in 1907, but the term 'dislocation' to refer to a defect on the atomic scale was coined by G. I. Taylor in 1934. Some types of dislocations can be visualized as being caused by the termination of a plane of atoms in the middle of a crystal. In such a case, the surrounding planes are not straight, but instead they bend around the edge of the terminating plane so that the crystal structure is perfectly ordered on either side. This phenomenon is analogous to the following situation related to a stack of paper: If half of a piece of paper is inserted into a stack of paper, the defect in the stack is noticeable only at the edge of the half sheet.

The two primary types of dislocations are *edge dislocations* and *screw dislocations*. *Mixed dislocations* are intermediate between these.

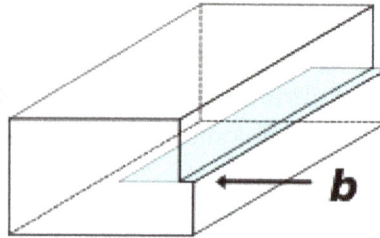

An edge-dislocation (b = Burgers vector)

Mathematically, dislocations are a type of topological defect, sometimes called a soliton. Dislocations behave as stable particles: they can move around, but maintain their identity. Two dislocations of opposite orientation can cancel when brought together, but a single dislocation typically cannot "disappear" on its own.

Geometry

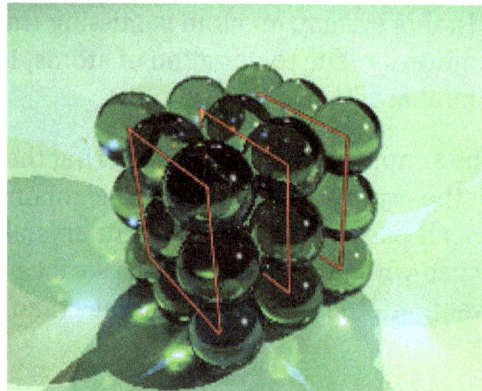

Crystal lattice showing atoms and lattice planes

Two main types of dislocations exist: edge and screw. Dislocations found in real materials are typically *mixed*, meaning that they have characteristics of both.

A crystalline material consists of a regular array of atoms, arranged into lattice planes (imagine stacking oranges in a grocery, each of the trays of oranges are the lattice planes). One approach is to begin by considering a 3D representation of a perfect crystal lattice, with the atoms represented by spheres. The viewer may then start to simplify the representation by visualising planes of atoms instead of the atoms themselves.

Edge

An edge dislocation is a defect where an extra half-plane of atoms is introduced midway through the crystal, distorting nearby planes of atoms. When enough force is applied from one side of the crystal structure, this extra plane passes through planes of atoms breaking and joining bonds with them until it reaches the grain boundary. A simple schematic diagram of such atomic planes can be used to illustrate lattice defects such as dislocations. The dislocation has two properties, a line direction, which is the direction running along the bottom of the extra half plane, and the Burgers vector which describes the magnitude and direction of distortion to the lattice. In an edge dislocation, the Burgers vector is perpendicular to the line direction.

The stresses caused by an edge dislocation are complex due to its inherent asymmetry. These stresses are described by three equations:

$$\sigma_{xx} = \frac{-\mu b}{2\pi(1-v)} \frac{y(3x^2 + y^2)}{(x^2 + y^2)^2}$$

$$\sigma_{yy} = \frac{\mu b}{2\pi(1-v)} \frac{y(x^2 - y^2)}{(x^2 + y^2)^2}$$

$$\tau_{xy} = \frac{\mu b}{2\pi(1-v)} \frac{x(x^2 - y^2)}{(x^2 + y^2)^2}$$

where μ is the shear modulus of the material, b is the Burgers vector, v is Poisson's ratio and x and y are coordinates.

These equations suggest a vertically oriented dumbbell of stresses surrounding the dislocation, with compression experienced by the atoms near the "extra" plane, and tension experienced by those atoms near the "missing" plane.

Screw

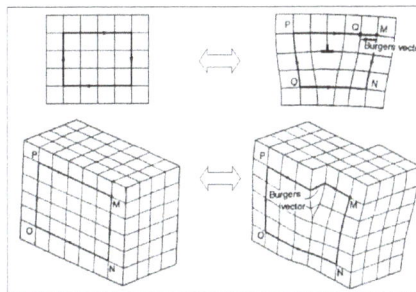

Top right: edge dislocation. Bottom right: screw dislocation.

Schematic diagram (lattice planes) showing a screw dislocation.

A *screw dislocation* is much harder to visualize. Imagine cutting a crystal along a plane and slipping one half across the other by a lattice vector, the halves fitting back together without leaving a defect. This is similar to the Riemann surface of the complex logarithm. If the cut only goes part way through the crystal, and then slipped, the boundary of the cut is a screw dislocation. It comprises a structure in which a helical path is traced around the linear defect (dislocation line) by the atomic planes in the crystal lattice. Perhaps the closest analogy is a spiral-sliced ham. In pure screw dislocations, the Burgers vector is parallel to the line direction.

Despite the difficulty in visualization, the stresses caused by a screw dislocation are less complex than those of an edge dislocation. These stresses need only one equation, as symmetry allows only one radial coordinate to be used:

$$\tau_r = \frac{-\mu b}{2\pi r}$$

where μ is the shear modulus of the material, b is the Burgers vector, and r is a radial coordinate. This equation suggests a long cylinder of stress radiating outward from the cylinder and decreasing with distance. Please note, this simple model results in an infinite value for the core of the dislocation at r=0 and so it is only valid for stresses outside of the core of the dislocation. If the Burgers vector is very large, the core may actually be empty resulting in a micropipe, as commonly observed in silicon carbide.

Mixed

In many materials, dislocations are found where the line direction and Burgers vector are neither perpendicular nor parallel and these dislocations are called *mixed dislocations*, consisting of both screw and edge character.

Partial

Dislocations can decompose into partial dislocations in order to facilitate movement through a crystal lattice.

Observation

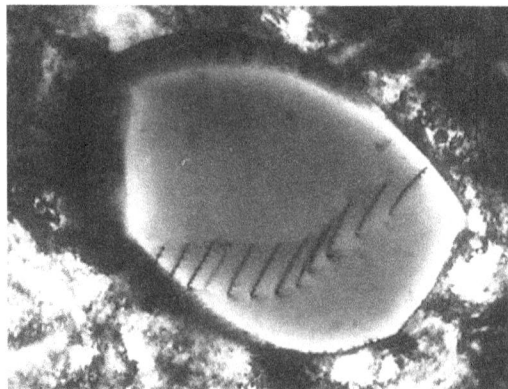

Transmission electron micrograph of dislocations

When a dislocation line intersects the surface of a metallic material, the associated strain field locally increases the relative susceptibility of the material to acidic etching and an etch pit of regular geometrical format results. If the material is strained (deformed) and repeatedly re-etched, a series of etch pits can be produced which effectively trace the movement of the dislocation in question.

Transmission Electron Microscopy (TEM)

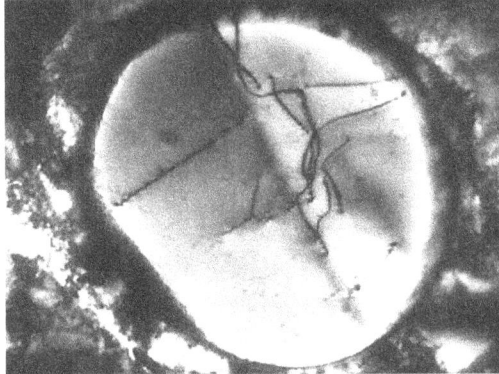

Transmission electron micrograph of dislocations

Transmission electron microscopy can be used to observe dislocations within the microstructure of the material. Thin foils of material are prepared to render them transparent to the electron beam of the microscope. The electron beam undergoes diffraction by the regular crystal lattice planes into a diffraction pattern and contrast is generated in the image by this diffraction (as well as by thickness variations, varying strain, and other mechanisms). Dislocations have different local atomic structure and produce a strain field, and therefore will cause the electrons in the microscope to scatter in different ways. Note the characteristic 'wiggly' contrast of the dislocation lines as they pass through the thickness of the material in the figure (also note that dislocations cannot end in a crystal, and these dislocations are terminating at the surfaces since the image is a 2D projection).

Dislocations do not have random structures, the local atomic structure of a dislocation is determined by the Burgers vector. One very useful application of the TEM in dislocation imaging is the ability to experimentally determine the Burgers vector. Determination of the Burgers vector is achieved by what is known as $\vec{g} \cdot \vec{b}$ ("g dot b") analysis. When performing dark field microscopy with the TEM, a diffracted spot is selected to form the image (as mentioned before, lattice planes diffract the beam into spots), and the image is formed using only electrons that were diffracted by the plane responsible for that diffraction spot. The vector in the diffraction pattern from the transmitted spot to the diffracted spot is the $\vec{g}$ vector. Without going into the finer points of electron microcopy; the contrast of a dislocation is scaled by a factor of the dot product of this vector and the Burgers vector ($\vec{g} \cdot \vec{b}$). As a result, if the burgers vector and $\vec{g}$ vector are perpendicular, there will be no signal from the dislocation and the dislocation will not appear at all in the image. Therefore, by examining different dark field images formed from spots with different g vectors, the burgers vector can be determined.

Also, some microscopes also permit the in-situ heating and/or deformation of samples, thereby permitting the direct observation of dislocation movement and their interactions.

Other Methods

Some microscopes also permit the in-situ heating and/or deformation of samples, thereby permitting the direct observation of dislocation movement and their interactions. Note the characteristic 'wiggly' contrast of the dislocation lines as they pass through the thickness of the material. Note also that a dislocation cannot end within a crystal; the dislocation lines in these images end at the sample surface. A dislocation can only be contained within a crystal as a complete loop.

Field ion microscopy and atom probe techniques offer methods of producing much higher magnifications (typically 3 million times and above) and permit the observation of dislocations at an atomic level. Where surface relief can be resolved to the level of an atomic step, screw dislocations appear as distinctive spiral features – thus revealing an important mechanism of crystal growth: where there is a surface step, atoms can more easily add to the crystal, and the surface step associated with a screw dislocation is never destroyed no matter how many atoms are added to it.

(By contrast, traditional optical microscopy, which is not appropriate for the *direct* observation of dislocations, typically offers magnifications up to a maximum of only around 2000 times).

After chemical etching, small pits are formed where the etching solution preferentially attacks the sample surface around the dislocations intercepting this surface, due to the more highly strained state of the material . Thus, the image features indicate points at which dislocations intercept the sample surface. In this way, dislocations in silicon, for example, can be observed *indirectly* using an interference microscope. Crystal orientation can be determined by the shape of the etch pits associated with the dislocations (in the case of the illustration below; 100 elliptical, 111 – triangular/pyramidal).

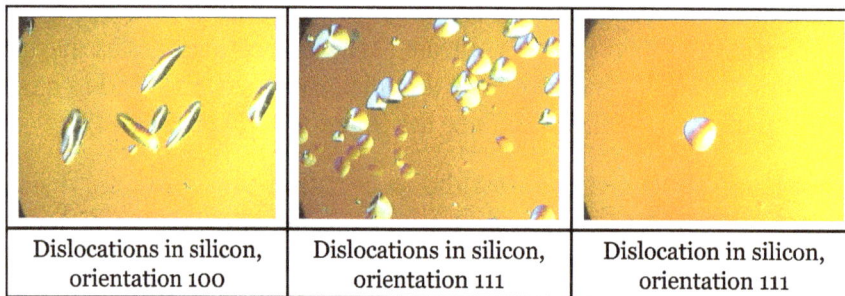

| Dislocations in silicon, orientation 100 | Dislocations in silicon, orientation 111 | Dislocation in silicon, orientation 111 |

Sources

Dislocation density in a material can be increased by plastic deformation by the following relationship: $\tau \propto \rho^{1/2}$. Since the dislocation density increases with plastic deformation, a mechanism for the creation of dislocations must be activated in the material. Three mechanisms for dislocation formation are homogeneous nucleation, grain boundary initiation, and interfaces between the lattice and the surface, precipitates, dispersed phases, or reinforcing fibers.

The creation of a dislocation by homogeneous nucleation is a result of the rupture of the atomic bonds along a line in the lattice. A plane in the lattice is sheared, resulting in 2 oppositely faced half planes or dislocations. These dislocations move away from each other through the lattice. Since homogeneous nucleation forms dislocations from perfect crystals and requires the simultaneous breaking of many bonds, the energy required for homogeneous nucleation is high. For instance, the stress required for homogeneous nucleation in copper has been shown to be $\dfrac{\tau_{hom}}{G} = 7.4 \times 10^{-2}$, where G is the shear modulus of copper (46 GPa). Solving for τ_{hom}, we see that the required stress is 3.4 GPa, which is very close to the theoretical strength of the crystal. Therefore, in conventional deformation homogeneous nucleation requires a concentrated stress, and is very unlikely. Grain boundary initiation and interface interaction are more common sources of dislocations.

Irregularities at the grain boundaries in materials can produce dislocations which propagate into the grain. The steps and ledges at the grain boundary are an important source of dislocations in the early stages of plastic deformation.

A well known source of dislocations by multiplication is the Frank-Read source.

The surface of a crystal can produce dislocations in the crystal. Due to the small steps on the surface of most crystals, stress in some regions on the surface is much larger than the average stress in the lattice. This stress leads to dislocations. The dislocations are then propagated into the lattice in the same manner as in grain boundary initiation. In single crystals, the majority of dislocations are formed at the surface. The dislocation density 200 micrometres into the surface of a material has been shown to be six times higher than the density in the bulk. However, in polycrystalline materials the surface sources cannot have a major effect because most grains are not in contact with the surface.

The interface between a metal and an oxide can greatly increase the number of dislocations created. The oxide layer puts the surface of the metal in tension because the oxygen atoms squeeze into the lattice, and the oxygen atoms are under compression. This greatly increases the stress on the surface of the metal and consequently the amount of dislocations formed at the surface. The increased amount of stress on the surface steps results in an increase in dislocations.

- The stresses produced by a dislocation source may be visualized by photoelasticity in a gamma-irradiated LiF single crystal. The tensile stress along the glide plane is red. The compressive stress is dark green:

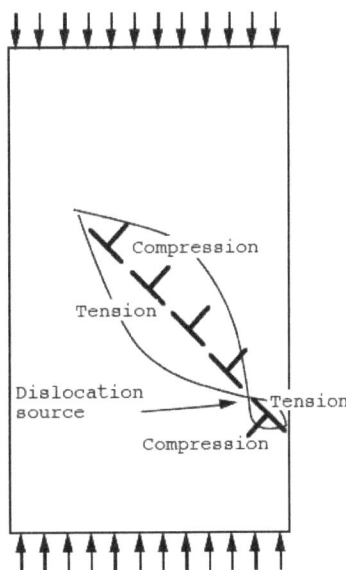

Dislocation source (schematic)

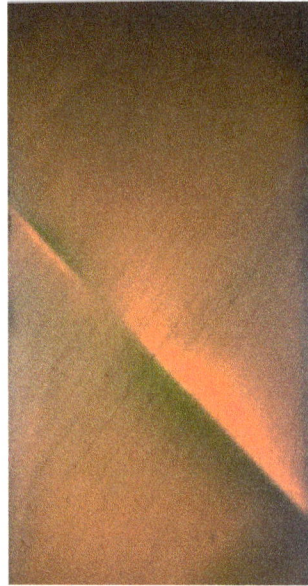

LiF irradiated with gamma. The slip plane is visible as a diagonal line, with compressive forces showing green and tensile forces showing red.

Slip and Plasticity

Until the 1930s, one of the enduring challenges of materials science was to explain plasticity in microscopic terms. A simplistic attempt to calculate the shear stress at which neighbouring atomic planes *slip* over each other in a perfect crystal suggests that, for a material with shear modulus G, shear strength τ_m is given approximately by:

$$\tau_m = \frac{G}{2\pi}.$$

As shear modulus in metals is typically within the range 20 000 to 150 000 MPa, this is difficult to reconcile with shear stresses in the range 0.5 to 10 MPa observed to produce plastic deformation in experiments.

In 1934, Egon Orowan, Michael Polanyi and G. I. Taylor, almost simultaneously realized that plastic deformation could be explained in terms of the theory of dislocations. Dislocations can move if the atoms from one of the surrounding planes break their bonds and rebond with the atoms at the terminating edge. In effect, a half plane of atoms is moved in response to shear stress by breaking and reforming a line of bonds, one (or a few) at a time. The energy required to break a single bond is far less than that required to break all the bonds on an entire plane of atoms at once. Even this simple model of the force required to move a dislocation shows that plasticity is possible at much lower stresses than in a perfect crystal. In many materials, particularly ductile materials, dislocations are the "carrier" of plastic deformation, and the energy required to move them is less than the energy required to fracture the material. Dislocations give rise to the characteristic malleability of metals.

When metals are subjected to "cold working" (deformation at temperatures which are relatively low as compared to the material's absolute melting temperature, T_m, i.e., typically less than 0.4 T_m)

the dislocation density increases due to the formation of new geometrically necessary dislocations and dislocation multiplication. The consequent increasing overlap between the strain fields of adjacent dislocations gradually increases the resistance to further dislocation motion. This causes a hardening of the metal as deformation progresses. This effect is known as strain hardening (also "work hardening"). Tangles of dislocations are found at the early stage of deformation and appear as non well-defined boundaries; the process of dynamic recovery leads eventually to the formation of a cellular structure containing boundaries with misorientation lower than 15° (low angle grain boundaries). In addition, adding pinning points that inhibit the motion of dislocations, such as alloying elements, can introduce stress fields that ultimately strengthen the material by requiring a higher applied stress to overcome the pinning stress and continue dislocation motion.

The effects of strain hardening by accumulation of dislocations and the grain structure formed at high strain can be removed by appropriate heat treatment (annealing) which promotes the recovery and subsequent recrystallisation of the material.

The combined processing techniques of work hardening and annealing allow for control over dislocation density, the degree of dislocation entanglement, and ultimately the yield strength of the material.

Climb

Dislocations can slip in planes containing both the dislocation line and the Burgers vector. For a screw dislocation, the dislocation line and the Burgers vector are parallel, so the dislocation may slip in any plane containing the dislocation. For an edge dislocation, the dislocation and the Burgers vector are perpendicular, so there is only one plane in which the dislocation can slip. There is an alternative mechanism of dislocation motion, fundamentally different from slip, that allows an edge dislocation to move out of its slip plane, known as dislocation climb. Dislocation climb allows an edge dislocation to move perpendicular to its slip plane. A creep mechanism involving only dislocation climb, also known as Harper-Dorn creep, can occur under certain conditions.

The driving force for dislocation climb is the movement of vacancies through a crystal lattice. If a vacancy moves next to the boundary of the extra half plane of atoms that forms an edge dislocation, the atom in the half plane closest to the vacancy can "jump" and fill the vacancy. This atom shift "moves" the vacancy in line with the half plane of atoms, causing a shift, or positive climb, of the dislocation. The process of a vacancy being absorbed at the boundary of a half plane of atoms, rather than created, is known as negative climb. Since dislocation climb results from individual atoms "jumping" into vacancies, climb occurs in single atom diameter increments.

During positive climb, the crystal shrinks in the direction perpendicular to the extra half plane of atoms because atoms are being removed from the half plane. Since negative climb involves an addition of atoms to the half plane, the crystal grows in the direction perpendicular to the half plane. Therefore, compressive stress in the direction perpendicular to the half plane promotes positive climb, while tensile stress promotes negative climb. This is one main difference between slip and climb, since slip is caused by only shear stress.

One additional difference between dislocation slip and climb is the temperature dependence. Climb occurs much more rapidly at high temperatures than low temperatures due to an increase in vacancy motion. Slip, on the other hand, has only a small dependence on temperature.

Stress Field of Dislocation

Dislocation represents a boundary between deformed and un-deformed parts of a crystal on the slip plane. Burgers vector b represents the magnitude of local displacement. As a result there will be severe lattice distortion around this. This is a measure of local strain (or stress) field. Very near the dislocation it is very high. However it drops rapidly with distance. It may be assumed to be elastic beyond a few atomic distances from the line defect. It is necessary to have some idea about the nature of the stress field in order to understand dislocation interactions with other disloca-tions or crystal defects. The expressions for stress fields of dislocation have been derived with the assumptions that the crystal is isotropic and the stress field is elastic beyond a few atomic spacing from the exact core of the dislocation. Let us try to derive these for screw dislocation from a simple physical concept.

Stress Field of a Screw Dislocation

Let us visualize a hollow cylinder. The internal radius ro represents the dimension of the disloca-tion core.

We would assume that the deformation within this is extremely large. Beyond this the stress field is elastic. Therefore it can be estimated using the theory of linear elasticity. Try to push the top half along the direction x_3 by a distance b. Now fix (weld) the two parts rigidly all along the plane of displacement. The process is exactly same as what happens when a screw dislocation is created in a crystal. The axis x_3 of the cylinder is the dislocation. The displacement b is its Burgers vector. The hole represents the dislocation core where the stress field is extremely large. Linear elastic theory is not applicable within this region. Note that as a result of the process there will be elastic strain field around the core. The area under the plot represents elastic stored energy. The partly sectioned cylinder has been deformed by movement on x1x3 plane against the direction x_3. The displacement along x_1 & x_2 are zero: $u_1 = u_2 = 0$. If you draw a clockwise circuit on the front face of the hollow cylinder you would have moved through a distance b against the direction x_3 on com-pletion of one rotation of magnitude 2π.

Assume that on application of shear stress τ on plane x_1x_3, the two parts of the partly sectioned cylinder has been moved through a distance b against the direction x_3. If the two parts are welded all along the sliced plane and the stress is withdrawn it will not come back to its initial state. Rather an internal stress would develop. This would represent the nature of the stress field around a screw dislocation.

Noting that the only non-zero component of displacement is $u_3 = -\dfrac{b}{2\pi}\theta = -\dfrac{b}{2\pi}\tan^{-1}\dfrac{x_2}{x_1}$; elas-tic strains are given by $\varepsilon_{ij} = 0.5\left(\dfrac{\partial u_i}{\partial x_j} + \dfrac{\partial u_j}{\partial x_i}\right)$ and $\sigma_{ij} = C_{ijkl}\varepsilon_{kl}$ it is possible derive the expressions

for stress field of a screw dislocation. Since u_3 is independent of x3 we do not expect any ax-ial strain or stress field in this case. The only non-zero components of strains / stresses are $\varepsilon_{13}, \varepsilon_{23}$ & σ_{13}, σ_{23}.

The screw dislocation does not have hydrostatic strain (stress) field. Therefore it does interact with point defects having only hydrostatic stress field. Also note that the stresses at any point in the neighborhood are inversely proportional to its distance from the dislocation. Its magnitude becomes extremely large as you approach the dislocation. It could be higher than the yield strength of the metal. No metal can support stresses of such high magnitude. The expressions for the stresses obtained as above may not be applicable at distances very close to the dislocation. However let us not bother about it. It is enough to know that a screw dislocation has a shear stress field around it. Its magnitude diminishes as you move away from it.

$$\varepsilon_{ij} = \left(\frac{\partial u_i}{\partial x_j} + \frac{\partial u_j}{\partial x_i} \right) / 2 \ \& \ u_1 = u_2 = 0, u_3 = -\frac{b\theta}{2\pi} = -\frac{b}{2\pi} \tan^{-1} \frac{x_2}{x_1}$$

$$\varepsilon_{11} = \varepsilon_{22} = \varepsilon_{33} = \varepsilon_{12} = 0 \ \therefore \ \varepsilon_{13} = \left(\frac{\partial u_3}{\partial x_1} \right) \ \& \ \varepsilon_{23} = \left(\frac{\partial u_3}{\partial x_2} \right)$$

$$\varepsilon_{ij} = \left(\frac{\partial u_i}{\partial x_j} + \frac{\partial u_j}{\partial x_i} \right) / 2 \ \& \ u_3 = \frac{b\theta}{2\pi} = \frac{b}{2\pi} \tan^{-1} \frac{x_2}{x_1}$$

$$\begin{pmatrix} 0 & 0 & \varepsilon_{13} \\ 0 & 0 & \varepsilon_{23} \\ \varepsilon_{13} & \varepsilon_{23} & 0 \end{pmatrix}$$

$$\varepsilon_{13} = \left(\frac{\partial u_3}{\partial x_1} \right) / 2 = \frac{b}{4\pi} \frac{x_2}{x_1^2 + x_2^2}$$

$$\varepsilon_{23} = \left(\frac{\partial u_3}{\partial x_2} \right) / 2 = -\frac{b}{4\pi} \frac{x_1}{x_1^2 + x_2^2}$$

Equation above illustrates how the strain field around a screw dislocation can be obtained from displacement gradients.

$$\begin{pmatrix} 0 & 0 & \sigma_{13} \\ 0 & 0 & \sigma_{23} \\ \sigma_{13} & \sigma_{23} & 0 \end{pmatrix}$$

$$\sigma_{13} = \frac{Gb}{2\pi} \frac{x_2}{x_1^2 + x_2^2}$$

$$\sigma_{23} = -\frac{Gb}{2\pi} \frac{x_1}{x_1^2 + x_2^2}$$

$$\sigma_{23} = -\frac{Gb}{2\pi} = \tau_{th} = \frac{G}{2\pi}$$

$$r_0 \approx b$$

This is obtained by multiplying the expression for elastic strain by shear modulus G. The expression could be used to estimate the magnitude of shear stress on the internal surface of the core. Assume $x_2 = 0$ & $x_2 = r_0$. This comes out to be of the order of $G / 2\pi$ which is nearly equal to the theoretical strength of metals.

The analysis shows that screw dislocation does not have hydrostatic strain (stress) field. Therefore it does interact with point defects having only hydrostatic stress field. Also note that the stresses at any point in the neighborhood are inversely proportional to its distance from the dislocation. Its magnitude becomes extremely large as you approach the dislocation. It could be higher than the yield strength of the metal. No metal can support stresses of such high magnitude. The expressions for the stresses obtained as above may not be applicable at distances very close to the dislocation. However let us not bother about it. It is enough to know that a screw dislocation has a shear stress field around it. Its magnitude diminishes as you move away from it.

Edge Dislocation – Stress Field

Derivation for the stress field is not as simple as in the case of a screw dislocation. However it easy to guess which of the three components of displacement vectors is zero. Due to shear stress acting on the sliced plane of a hollow cylinder and subsequent welding a situation similar to that in the case of an edge dislocation can arise. Note that in this case at a distance r from the dislocation the displacement along the slip plane against the direction x_1 is b. But it is a function of distance of point on the slip plane from the dislocation (axis of the cylinder). Also note that although $u_2 = 0$ for any point on the slip plane, but it has a definite value depending on its coordinates as you move up or down the slip plane. Both of these are independent of x_3. Apart from this there is no displacement along the axial direction x3. This means $u_3 = 0$. This is true for any point within the cylinder. Therefore it is a classic example of a plane strain situation. The non-zero components of strains & stresses are $\{\varepsilon_{11}\varepsilon_{12}\varepsilon_{13}\}$ and $\{\sigma_{11}\sigma_{12}\sigma_{13}\sigma_{33}\}$ respectively.

$$\begin{pmatrix} \sigma_{11} & \sigma_{12} & 0 \\ \sigma_{12} & \sigma_{22} & 0 \\ 0 & 0 & \sigma_{33} \end{pmatrix}$$

$$\sigma_{11} = -\frac{Gb}{2\pi(1-v)}\frac{x_2(3x_1^2+x_2^2)}{(x_1^2+x_2^2)^2}$$

$$\sigma_{22} = \frac{Gb}{2\pi(1-v)}\frac{x_2(x_1^2-x_2^2)}{(x_1^2+x_2^2)^2}$$

$$\sigma_{12} = \frac{Gb}{2\pi(1-v)}\frac{x_1(x_1^2-x_2^2)}{(x_1^2+x_2^2)^2}$$

$$\sigma_{33} = v(\sigma_{11}+\sigma_{22})$$

It has undergone slip due to shear on the slip plane. After slip the two parts are welded. This denotes all the features of deformation around an edge dislocation. The axis of the cylinder is along x3. The slip plane is x_1x_3. The magnitude of displacement on the slip plane at the edge is b. Note:

$$u_1 = f(b,x_1,x_2); u_2 = f(b,x_1,x_2); \& u_3 = 0$$

It may be seen that the slip plane is represented by $x_2 = 0$. The only stress acting on this plane is given by

$$\sigma_{12} = \frac{Gb}{2\pi(1-v)x_1}$$

Also note that above slip plane σ_{11} is compressive. Its magnitude is much larger than σ_{22}. This also signifies that the stress field is compressive above the slip plane whereas it is tensile beneath it. From the expressions it is evident that edge dislocation has a strong hydrostatic stress field. This is given by $\sigma_{hydro} = \frac{1}{3}(\sigma_{11} + \sigma_{22} + \sigma_{33})$. Therefore unlike screw dislocation it can also interact with symmetric point defects having only hydrostatic stress field.

Elastic Stored Energy of Dislocation

Using the expressions for stress field it is possible to find the magnitude of stored elastic energy. Note that the energy per unit volume is given by $E = \int \sigma_{ij} d\varepsilon_{ij}$. Note that the use of repeated suffix denotes summation. The limits of integration should cover the entire region from the core to the outer surface. In the case of screw dislocation derivation is easy to follow. It can be further simplified using cylindrical coordinate system. In this case stress field is axi-symmetric. It depends only on the distance r from the axis.

$$E = \int_{r_0}^{R} G \frac{\gamma^2}{2} 2\pi rL dr = \int_{r_0}^{R} \pi GL \left(\frac{b}{2\pi r}\right)^2 r dr = \frac{LGb^2}{4\pi} \ln\left(\frac{R}{r_0}\right)$$

$$E_{core} = \frac{\tau\gamma}{2} V = \frac{\tau^2}{2G} V \approx \left(\frac{G}{2\pi}\right)^2 \frac{\pi r_0^2 L}{2G} \approx \frac{Gb^2}{8\pi} L$$

The energy of an edge dislocation can also be estimated using the expressions for stress fields. The expressions are exactly similar. The energy for both edge and screw dislocation is proportional to Gb^2. So far we have ignored the energy store in the core of the dislocation. It is true that the stress fields within this zone do not follow linear elastic behavior. However it is still possible to guess its magnitude. A mixed dislocation has both edge and screw components. Its energy can be estimated by substituting these in respective equations.

$$E = (E_e + E_{core})/L$$

$$E^s = \frac{Gb^2}{4\pi}\left(\ln\frac{r}{r_0}+1\right) \qquad r_0 = b = 0.3nm \ \& \ r = 30\mu m$$

$$E^e = \frac{Gb^2}{4\pi(1-v)}\left(\ln\frac{r}{r_0}+1\right) \qquad (r/r_0) = 10^5 \ \& \ \ln(r/r_0) \sim 11$$

$$E^d = E^s = E^e \sim 0.5Gb^2$$

The above equation gives the expressions for energy per unit length of an edge & a screw dislocation. Note that the difference between the size r of the crystal and the radius of the core ro is large. Therefore energy within the core is expected to be much less. It is taken to be of the order of $\frac{Gb^2}{4\pi}$. The energy of edge dislocation includes a term $(1-v)$ in the denominator where v is the Poisson ratio. Since its value is 0.3. The energy of an edge dislocation is a little higher than that of a screw dislocation.

Note that the difference between the energy of an edge & a screw dislocation is marginal. Therefore irrespective of the type of dislocation its energy can be taken as $0.5Gb^2$. G is the shear modulus of the crystal. If you look the expression for elastic stored energy of a dislocation it may appear that it can approach infinity (be extremely large). However it does not happen because stress field becomes negligible beyond a certain distance. It may be assumed to extend up to a distance of around $1000r_0$.

Force on a Dislocation

Dislocation can glide on a slip plane if the resolved stress on the plane exceeds a critical value called CRSS (critical resolved shear stress). This may be visualized that the movement of dislocation is due to a force F acting on it.

Dislocation Vacancy Interaction

The presence of a point defect is associated with local lattice distortion. In the case of a vacancy it is the same in all directions. Therefore the nature of the stress field is hydrostatic. It can only interact with edge dislocation. It helps dislocation to move in a direction perpendicular to its slip plane. This is known as a non-conservative motion of dislocation. When a vacancy moves to an edge dislocation it climbs up where as if an atom moves to a point beneath it climbs down.

Glide is the most common mode of dislocation movement. Both screw & edge dislocation can glide. However edge dislocation can also move in a direction normal to its slip plane. If a vacancy from the crystal lattice moves to position just beneath the extra plane of atom then it moves up. The plane on which it could glide is different. As a result vacancy concentration decreases.

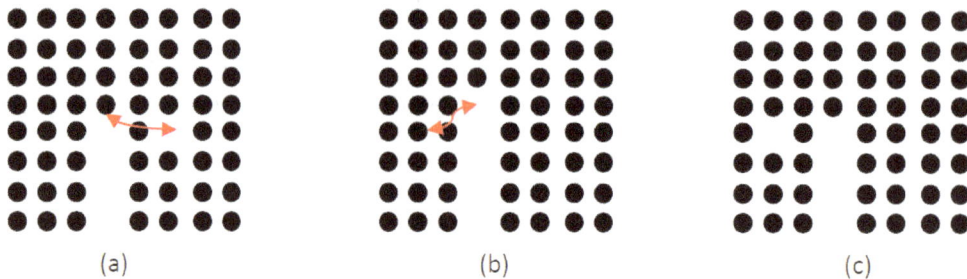

(a)   (b)   (c)

Figure illustrates interaction between vacancy and edge dislocation.

(a) Note the initial location of a dislocation and a vacancy. When the vacancy moves to a site occupied by an atom as shown by the arrow the dislocation climbs up. This is shown in (b). The site beneath the extra half plane may be considered as a vacant site. Exchange of position as shown by the red arrow in.

(b) Helps the dislocation climb down.

(c) Shows that a vacant site must be created when a dislocation climbs down.

Force Acting on a Dislocation

The resolved shear stress on a plane is τ; the force on the dislocation is τb; where b denotes Burgers vector. However; often the local stress field around a defect is given in the form of a stress matrix (or stress tensor σ_{ij}). In such a situation we need a more generalized approach to find out the force acting on a dislocation. It has to be a function of σ, b & t. Note that τb denotes a product of stress and Burgers vector where τ is one of the components of a stress tensor. The product is a vector having magnitude and direction. If the local stress field is given by σ_{ij}, a tensor of second rank, the force on the dislocation should be given by $\overline{\overline{\sigma}}.\overline{b}$. The double bar is a sign of second rank tensor whereas a single bar denotes a vector. This truly represents a product of 3x3 matrix with 3x1 vector giving a 3x1 vector. Its vector product with the unit vector along the dislocation direction at a point represents the force acting on it at that point. This can be mathematically represented as follows:

$$\overline{F}=\overline{\overline{\sigma}}\,\overline{b}\times\overline{t}\text{ or; } F_i =\epsilon_{ijk}\,\sigma_{jl}b_l t_k \tag{1}$$

Note that the both the forms of the equation are the same. The former is in vector notation whereas the latter is in tensor notation. The use of repeated suffix represents summation. For example the subscript l 'l' in equation 1 appears in σ_{il} and b_l. This means the suffix l can have values from 1 to 3. More explicitly this is given by:

$$\sigma_{il}b_l = \sigma_{i1}b_1 +\sigma_{i2}b_2 +\sigma_{i3}b_3 \tag{2}$$

Physically this represents the component of the vector along the axis i. Using this we could get the magnitude of the vector along the three orthogonal axes. The force on the dislocation is given by its cross product with vector t representing dislocation direction. The symbol ϵ_{ijk} is a coefficient given by the following equation:

$$\epsilon_{123}=\epsilon_{231}=\epsilon_{321}= +1 \text{ \& } \epsilon_{132}=\epsilon_{321}=\epsilon_{213}=-1 \tag{3}$$

$$\epsilon_{ijk} = 0 \text{ if any two or more of the subscripts are the same} \tag{4}$$

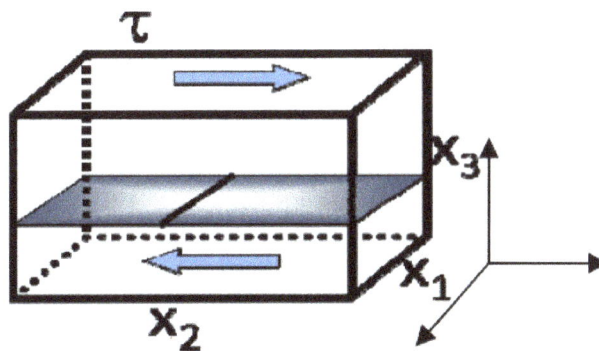

A slip plane with a dislocation (the dark line on the shaded plane).

The slip plane is perpendicular to x_3 axis. Shear stress τ acts along the direction x_2 on the top face of the crystal.

If the crystal is loaded as shown the only non-zero component of the stress is $\tau = \sigma_{23} = \sigma_{32}$. Note that the dislocation is aligned along x_1 axis. Therefore the nonzero component of t representing dislocation direction is t_1. If we assume this to be an edge dislocation the only non zero component of Burgers vector b is b_2. Therefore the force on the dislocation is given by

$$F_i = \epsilon_{ijl}\, \sigma_{j2} b_2 t_1 = \epsilon_{i31}\, \sigma_{32} b_2 t_1 \tag{5}$$

Thus the only non-zero component of the force is $F_2 = \epsilon_{231}\, \sigma_{32} b_2 t_1 = \tau b$. In this case the derivation looks complex. However this does help understand the effect of any arbitrary state of stress on a dislocation.

Interaction between Two Parallel Screw Dislocations

The equation 1 can be used to find the force exerted by a screw dislocation located along the axis x_3 on another located at a distance r from it. Figure below shows the locations of the two parallel screw dislocations with respect to the reference axes $x_1 x_2 x_3$.

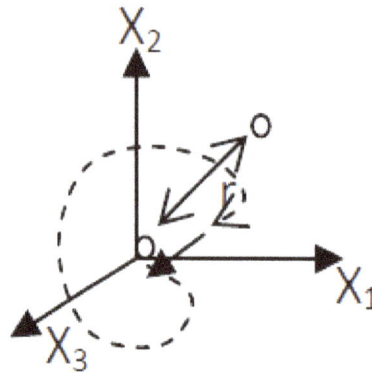

A figure showing locations of two screw dislocations lying along X3 axis. A clockwise circuit has been drawn in dotted line while looking against the direction x_3. The arrow to complete the circuit is the Burgers vector of the dislocation. Note dislocation & burgers vector are aligned in opposite direction.

The only non-zero components of stress field of a screw dislocation are σ_{13} & σ_{23}. The Burgers vector & the direction of the dislocation at a distance r are [o o b'] & [o o -1]; note that only the components along x3 are non-zero. Therefore the stress acting on the second dislocation due to the stress field of the first is given by

$$F_i = \epsilon_{ij3}\, \sigma_{j3} b_3 t_3 \tag{6}$$

Note that the suffix i & j must have different values. Thus

$$F_i = \epsilon_{123}\, \sigma_{23} b_3 t_3 \quad \& \quad F_2 = \epsilon_{213}\, \sigma_{13} b_3 t_3 \tag{7}$$

Recall that if b is the Burgers vector of the dislocation along x_3. The two of its non-zero stress components are

$$\sigma_{13} = \frac{Gb}{2\pi}\frac{x_2}{x_1^2 + x_2^2} \quad \& \quad \sigma_{23} = \frac{Gb'}{2\pi}\frac{x_2}{x_1^2 + x_2^2} \tag{8}$$

Substitute these in equation 8 and put $t_3 = -1$ & $b_3 = b'$ to get

$$F_1 = \frac{Gbb'}{2\pi}\frac{x_2}{x_1^2 + x_2^2} \;\&\; F_2 = \frac{Gbb'}{2\pi}\frac{x_2}{x_1^2 + x_2^2} \tag{9}$$

Using cylindrical coordinate you can get a still simpler expression for the forces. This gives

$$F_1 = \frac{Gbb'}{2\pi r}\cos\theta \;\&\; F_2 = \frac{Gbb'}{2\pi r}\sin\theta.$$

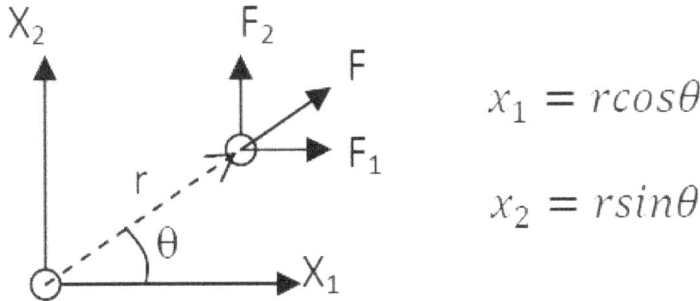

$$x_1 = r\cos\theta$$

$$x_2 = r\sin\theta$$

Figure showing the locations of dislocations. The distance between the two is r whereas the angle between r & X_1 axis is θ. The relations between the coordinates are also displayed.

The resultant force is thus $F = \dfrac{Gbb'}{2\pi r}$. Note its direction. In this case it is repulsive. There will be similar repulsive force acting on the other dislocation. As a result the two would move away from each other. In other words like dislocations repel each other. If the dislocations are of opposite nature they attract each other.

Parallel Edge Dislocation Having Parallel Burgers Vector

Following the steps described in the previous case you could easily find the forces acting on an edge dislocation due the stress field on another edge dislocation.

$[\sigma]$ denotes the stress field of the dislocation at the origin of the reference axes. Let b' be its Burgers vector. [b] & [t] denote the Burgers vector & dislocation direction of the second dislocation. The two dislocations are aligned along the axis x_3. Note the derivation steps. F_1 & F_2 are the two components of the force between the two dislocations. The magnitudes of these depend on x_1 & x_2.

$$\begin{pmatrix} \sigma_{11} & \sigma_{12} & 0 \\ \sigma_{12} & \sigma_{22} & 0 \\ 0 & 0 & \sigma_{33} \end{pmatrix} \quad [b] = \begin{bmatrix} b \\ 0 \\ 0 \end{bmatrix} \quad [t] = \begin{bmatrix} 0 \\ 0 \\ 1 \end{bmatrix}$$

$$F_i = \varepsilon_{ijk}\sigma_{jl}b_l t_k = \varepsilon_{ij3}\sigma_{jl}b_1 t_3 = \varepsilon_{i13}\sigma_{11}b_1 t_3 + \varepsilon_{i23}\sigma_{21}b_1 t_3$$

$$F_1 = \varepsilon_{123}\sigma_{21}b_1 t_3 = \frac{Gbb'}{2\pi(1-v)}\frac{x_1(x_1^2 - x_2^2)}{(x_1^2 + x_2^2)^2} \qquad \text{Like dislocations repel \& unlike}$$

$$F_2 = \varepsilon_{213}\sigma_{11}b_1t_3 = \frac{Gbb'}{2\pi(1-v)}\frac{x_1(3x_1^2+x_2^2)}{(x_1^2+x_2^2)^2} \qquad \text{dislocations attract}$$

x_1 & x_2 are the coordinates of the point of intersection of the second dislocation with plane $x_3 = 0$. If $x_2 = 0$ the two dislocations lie on the same plane & $F_2 = 0$. The sign of F_1 denotes that the two would repel each other. In short two like dislocation on the same plane would repel whereas unlike dislocations would attract each other. The direction and magnitudes of F_1 & F_2, the two components of the force are functions of the coordinates x_1 & x_2. Let us substitute $x_1 = r\cos\theta$ and $x_2 = r\sin\theta$ where r is the distance between the dislocations and θ is the angle between vector r and the axis x_1.

$$F_1 = \frac{Gbb'}{2\pi(1-v)}\frac{x_1(x_1^2-x_2^2)}{(x_1^2+x_2^2)^2} = \frac{Gbb'}{2\pi(1-v)r}\cos\theta\cos 2\theta$$

$$F_2 = \frac{Gbb'}{2\pi(1-v)}\frac{x_1(3x_1^2+x_2^2)}{(x_1^2+x_2^2)^2} = \frac{Gbb'}{2\pi(1-v)r}\sin\theta(1+2\cos^2\theta)$$

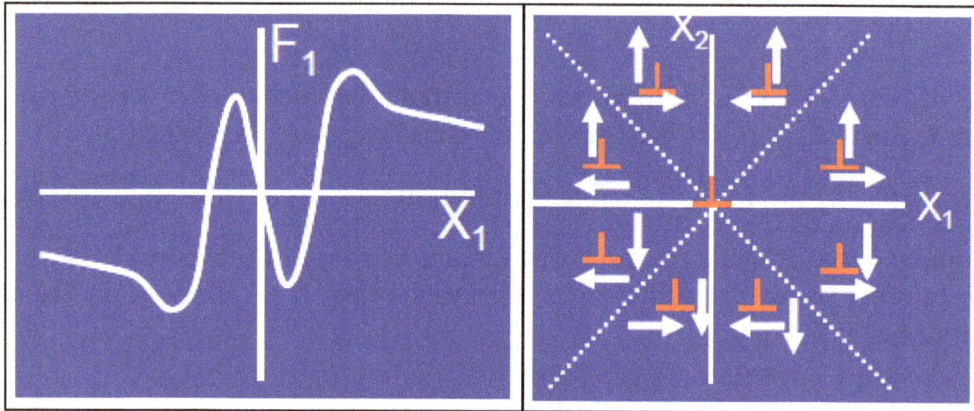

Figure above gives the expressions for the forces of interaction between two edge parallel edge dislocations; in both Cartesian and cylindrical coordinates. The sketch on the right denotes the locations of dislocations. There is one at the origin. The others are at various places on the 4 quadrants. The arrows denote the directions in which the forces act on these due to the stress field of the dislocation at the origin. Note the directions of F_2 above and below the plane $x_2 = 0$.

Above equation shows the directions of the force F_2 on the second dislocation depending on its locations. If it is above the slip plane of the first (the one at the origin) it would tend to climb up; whereas if it is below the slip plane it would tend to climb down. The dotted lines represent positions where the force $F_1 = 0$. Note that these are at 45° (It follows from the expressions given in polar coordinates). The directions along which F_1 acts on either side of these lines are different. The points on the dotted lines represent positions of unstable equilibrium. This gives a plot of force F_1 as a function of the distance x_1 on a plane $x_2 = $ constant. The location $x_1 = 0$ denotes a position of stable equilibrium. If the dislocation is forced to move a little on either of the two sides the force F_1 will help it come back to its initial position. At locations $x_1 = \pm x_2$ too $F_1 = 0$. In this case if it moves a little the force F_1 would tend to move it away from its position of unstable equilibrium. The most stable configuration of a set of parallel edge dislocations with parallel burgers vector is therefore one over the other.

Partial Dislocation

Partial dislocations are a decomposed form of dislocations that occur within a material.

Reaction Favorability

A dislocation will decompose into partial dislocations if the energy state of the sum of the partials is less than the energy state of the original dislocation. This is summarized by *Frank's Energy Criterion*:

$$|\mathbf{b}_1|^2 > |\mathbf{b}_2|^2 + |\mathbf{b}_3|^2 \quad \text{(favorable, will decompose)}$$
$$|\mathbf{b}_1|^2 < |\mathbf{b}_2|^2 + |\mathbf{b}_3|^2 \quad \text{(not favorable, will not decompose)}$$
$$|\mathbf{b}_1|^2 = |\mathbf{b}_2|^2 + |\mathbf{b}_3|^2 \quad \text{(will remain in original state)}$$

Shockley Partial Dislocations

Shockley partial dislocations generally refer to a pair of dislocations which can lead to the presence of stacking faults. This pair of partial dislocations can enable dislocation motion by allowing an alternate path for atomic motion.

$$\mathbf{b}_1 \rightarrow \mathbf{b}_2 + \mathbf{b}_3$$

In FCC systems, an example of Shockley decomposition is:

$$\frac{a}{2}[10\bar{1}] \rightarrow \frac{a}{6}[2\bar{1}\bar{1}] + \frac{a}{6}[11\bar{2}]$$

Which is energetically favorable:

$$|\frac{a}{2}\sqrt{1^2+0^2+(-1)^2}|^2 > |\frac{a}{6}\sqrt{2^2+(-1)^2+(-1)^2}|^2 + |\frac{a}{6}\sqrt{1^2+1^2+(-2)^2}|^2$$

$$\frac{a^2}{2} > \frac{a^2}{6} + \frac{a^2}{6}$$

The components of the *Shockley Partials* must add up to the original vector that is being decomposed:

$$\frac{a}{2}(1) = \frac{a}{6}(2) + \frac{a}{6}(1)$$
$$\frac{a}{2}(0) = \frac{a}{6}(-1) + \frac{a}{6}(1)$$
$$\frac{a}{2}(-1) = \frac{a}{6}(-1) + \frac{a}{6}(-2)$$

Frank Partial Dislocations

Frank partial dislocations are sessile, or immobile, but can move by diffusion of atoms. In FCC systems, Frank partials are given by:

$$\mathbf{b}_{frank} = \frac{a}{3}[1\ 1\ 1]$$

Thompson Tetrahedron

Shockley partials and Frank partials can combine to form a *Thompson tetrahedron*, or a *stacking fault tetrahedron*.

Lomer–Cottrell Lock

The Lomer–Cottrell lock is formed by partial dislocations and is sessile.

References

- Donald E. Sands (1994). "§4-2 Screw axes and §4-3 Glide planes". Introduction to Crystallography (Reprint of WA Benjamin corrected 1975 ed.). Courier-Dover. pp. 70–71. ISBN 0486678393

- Lieb, Klaus-Peter; Keinonen, Juhani (2006). "Luminescence of ion-irradiated α-quartz". Contemporary Physics. 47 (5): 305–331. Bibcode:2006ConPh..47..305L. doi:10.1080/00107510601088156

- Bernal, J. D.; Fowler, R. H. (1933). "A Theory of Water and Ionic Solution, with Particular Reference to Hydrogen and Hydroxyl Ions". The Journal of Chemical Physics. 1 (8): 515. Bibcode:1933JChPh...1..515B. doi:10.1063/1.1749327

- Schwartz, Mel (2002). "Tin and Alloys, Properties". Encyclopedia of Materials, Parts and Finishes (2nd ed.). CRC Press. ISBN 1-56676-661-3

- Spence, J. C. H.; et al. (2006). "Imaging dislocation cores – the way forward". Philos. Mag. 86: 4781. Bibcode:2006PMag...86.4781S. doi:10.1080/14786430600776322

- Pauling, Linus (1947). "Atomic Radii and Interatomic Distances in Metals". Journal of the American Chemical Society. 69 (3): 542.doi:10.1021/ja01195a024

- Mermin, N. (1979). "The topological theory of defects in ordered media". Reviews of Modern Physics. 51 (3): 591–648. Bibcode:1979RvMP...51..591M. doi:10.1103/RevModPhys.51.591

- Holleman, Arnold F.; Wiberg, Egon; Wiberg, Nils (1985). "Tin". Lehrbuch der Anorganischen Chemie (in German) (91–100 ed.). Walter de Gruyter. pp. 793–800. ISBN 3-11-007511-3

- Siegel, R. W. (1978). "Vacancy concentrations in metals". Journal of Nuclear Materials. 69-70: 117–146. Bibcode:1978JNuM...69..117S. doi:10.1016/0022-3115(78)90240-4

- James Shackelford (2009). Introduction to Materials Science for Engineers (7th ed.). Upper Saddle River, NJ 07458: Pearson Prentice Hall. pp. 110–11. ISBN 0-13-601260-4

- L. Pauling (1929). "The principles determining the structure of complex ionic crystals". J. Am. Chem. Soc. 51 (4): 1010–1026. doi:10.1021/ja01379a006

Permissions

Index

www.ingramcontent.com/pod-product-compliance
Lightning Source LLC
Chambersburg PA
CBHW082017190326
41458CB00010B/3214